QUANTUM MECHANICS ON MATTER PHYSICS

BEYOND THE ATOMIC LEVEL

ARYAN VINOD

Copyright © 2024 by Aryan Vinod

All rights reserved. No part of this book may be reproduced, distributed, or transmitted in any form or by any means, including photocopying, recording, or other electronic or mechanical methods, without the prior written permission of the author, except in the case of brief quotations embodied in critical reviews or articles.

ISBN 9798304584289

Cover design by Aryan Vinod

This is a work of non-fiction. While every effort has been made to ensure the accuracy of the information presented, the content of this book is for informational purposes only. The views expressed are those of the author and do not necessarily reflect the views of any institutions, organizations, or individuals mentioned in the text.

This book is dedicated to my friend's little brother Muhammad Ramzan and all physics lovers and young philosophers

CONTENTS

	Introduction	i
1	The birth of a new science	3
2	Are We Waves or Particles?	13
3	The Ghost in the Atom	21
4	The Limits of Knowing	41
5	Can Something Be Nothing?	51
6	Are We All Connected?	65
7	Does Time Really Exist?	95
8	Who Observes the Observer?	125

INTRODUCTION

Physics has viewed the atomic level as the proper foundation of our description of matter. For many centuries, atoms were considered to be indivisible, indestructible units and the fundamental building blocks of matter. This perception was supported by classical physics as the atoms were considered indestructible due to the extremely small radius of their core. In the late 19th and early 20th centuries, however, several phenomena were observed which could not be explained using classical theories. Some of those included blackbody radiation and the photoelectric effect. These difficulties finally led to the development of a new framework: quantum mechanics—a branch that unveiled the probabilistic and uncertain nature of the universe.

In the quantum regime, the rigid determinism of classical physics does not hold. The behaviour of matter is different and is controlled by laws that are mostly counterintuitive and incomprehensible. While great scientists have made remarkable contributions in this arena, its mysteries remain as big a challenge to the wit of man regarding the explanation of nature.

This work will explore that domain beyond the atom, where certainty is no longer in charge, but rather probabilities with intrinsic uncertainties. The aim is to give a basic understanding at an introductory level of the principles of quantum mechanics while trying to answer the resultant philosophically involved questions from the concept. Not being a university or postgraduate student, but only a 17-year-old who has relied on my mother's undergraduate textbooks and online research papers to compile this work, it may not be advanced but is intended to present the foundational ideas of quantum mechanics in a clear and accessible manner.

The study of quantum mechanics is the study of not only the tiniest building blocks of the universe but also of one more thing—questioning the nature of existence as we think we know it. What does it say about a universe that's run with the help of probabilities? What happens to our idea of reality if certainty at the very fundamental level is replaced by uncertainty? The purpose of this book is to bring in an ordered

BEYOND THE ATOMIC LEVEL

introduction to quantum mechanics that should seriously question your reality and your very existence.

1: THE BIRTH OF A NEW SCIENCE

When I was too young, I made up a theory about the revolution of Earth around the Sun. I said it's due to the magnetic field, which is attractive to every planet, and also there is an imaginary repulsive force which makes the Earth and other planets feel an equilibrium force, and then it turned into rotation. My teacher just smiled. After two years, I started knowing some bases of relativity, then I realized my stupidity. And when I went to higher classes, my teacher told me that centripetal force made Earth revolve around the Sun, and I was still confused. When I started knowing about quantum mechanics, he said gravity is made of gravitons. After that, my perspective about physics changed. Physics won't work the same in every scale. Something changed in some part, and some of them became nonexistent. Like that, all these laws of classical physics started breaking when explaining blackbody radiation.

A blackbody is an idealized object which, when subjected to incident electromagnetic radiation, at whatever wavelength or angle, will absorb that radiation completely. In turn, it will emit radiation according to a spectrum dependent only on the temperature of that body. This was codified in 1859 by Gustav Kirchhoff, who assumed that the radiation of a blackbody is a universal function depending only on the temperature and independent of the material properties of the body. Experimental work on blackbody cavities in the late 19th century provided an empirical determination of the spectral distribution of radiation, which showed features not understandable within classical physics. In particular, formulas such as the Rayleigh-Jeans law accurately described that part of the spectrum which falls in the region of longer wavelengths but predicted infinite energy output at shorter wavelengths—the so-called ultraviolet catastrophe. These failures indicated that a new theoretical approach was required, and this came with Max Planck in the year 1900. Planck introduced the revolutionary idea that energy is quantized, proposing that it is emitted or absorbed in discrete packets called quanta. It was a hypothesis which not only resolved the classical predictions, but it laid the foundation for quantum mechanics to completely change our thinking about energy, light, and matter.

In 1899, Lummer and Pringsheim conducted groundbreaking experiments to measure the spectral distribution of black body radiation with unprecedented accuracy. By using an improved black body cavity and precise instrumentation, they systematically recorded how the intensity of radiation varied with wavelength at different temperatures. Their results confirmed the existence of a peak in the radiation spectrum, which shifted to shorter wavelengths as temperature increased, in accordance with Wien's Displacement Law. These findings not only challenged classical physics but also provided the experimental foundation for Max Planck's revolutionary theory of quantized energy.

BEYOND THE ATOMIC LEVEL

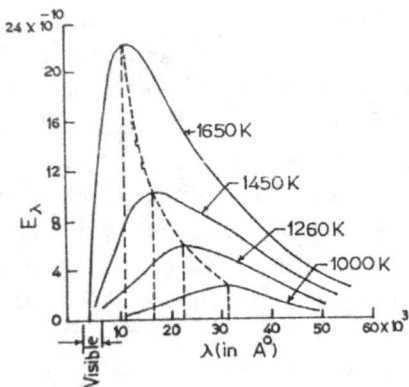

The spectral emissive power $E\lambda$ of a black body, defined as the energy radiated per unit area, per unit wavelength, per unit time, is expressed as:

$$E\lambda = \frac{d\lambda}{dE}$$

This quantity describes how the energy emitted by a black body is distributed across different wavelengths for a given temperature. The graph of black body radiation reveals several important conclusions. 1) As the temperature of the black body increases, the wavelength at which the intensity is highest shifts towards shorter wavelengths, in accordance with Wien's Displacement Law. This can be represented as:

$$\lambda_{max} = \frac{b}{T}$$

where λ_{max} is the peak wavelength, b is Wien's constant, and T is the temperature. 2) The total intensity of radiation emitted by the black body increases with temperature, as described by the Stefan-Boltzmann Law, which states that total energy radiated is proportional to the fourth power of the temperature:

$$I_{total} = \sigma T^4$$

where I_{total} is the total intensity, σ is the Stefan-Boltzmann constant, and T is the temperature. 3) The energy distribution of the radiation shifts towards shorter wavelengths and higher energies with increasing temperature, explaining why hotter objects appear bluer and cooler objects appear redder or infrared. 4) The graph also highlights the historical issue of the ultraviolet catastrophe, where classical physics predicted infinite radiation at short wavelengths. This issue is resolved by a more accurate model, which correctly predicts the intensity distribution without divergence.

5) Finally, the continuous nature of the radiation across all wavelengths is evident, supporting the idea that a black body absorbs and emits radiation perfectly across the entire electromagnetic spectrum.

Wien's law provides crucial insights into the behaviour of blackbody radiation and serves as a foundation for understanding stellar spectra, enabling astronomers to estimate the temperatures of stars based on their color.

Following Wien's Displacement Law, the Rayleigh-Jeans Law extends the study of blackbody radiation by describing the spectral energy distribution at long wavelengths (or low frequencies). Specifically, it provides an approximation for the intensity of electromagnetic radiation emitted by a blackbody at a given temperature. This law is pivotal in the history of physics, as it highlights the limitations of classical mechanics and led to the development of quantum mechanics. In the classical approach, Rayleigh and Jeans derived an expression for the intensity of radiation as a function of wavelength and absolute temperature :

$$I(\lambda, T) = \frac{2ck_B T}{\lambda^4}$$

where $I(\lambda, T)$ represents the intensity of radiation per unit wavelength. The Rayleigh-Jeans law is based on several key assumptions. Firstly, it assumes classical electromagnetic theory, treating the electromagnetic radiation inside the blackbody cavity as standing waves. Secondly, it relies on the equipartition of energy principle, where energy is evenly distributed across all available modes of radiation. Each mode contributes to the total energy. Finally, it accounts for the density of states, where the number of allowed wave modes per unit volume and per unit wavelength is proportional to the square of the wavelength's reciprocal. Using these assumptions, Rayleigh and Jeans calculated the total energy density of radiation within the cavity and obtained the above equation.

The behaviour of the Rayleigh-Jeans law can be analyzed in two regimes. At long wavelengths (low frequencies), the law matches experimental results and accurately predicts the intensity of radiation. For long wavelengths, the term dominates, leading to a gradual decrease in intensity. However, at short wavelengths (high frequencies), the law predicts that the intensity increases without bound as approaches zero. This leads to the so-called "ultraviolet catastrophe", where the predicted energy diverges, which is physically unrealistic. Experimentally, blackbody radiation does not behave this way; instead, it peaks and then drops off at shorter wavelengths.

The ultraviolet catastrophe refers to the failure of the Rayleigh-Jeans law to describe blackbody radiation at high frequencies (short wavelengths). As , the predicted intensity approaches infinity:

$$I(\nu, T) = \frac{8\pi\nu^2}{c^3} k_B T$$

This divergence was a major issue in classical physics because it implied infinite energy emission from blackbodies, which is nonsensical and contradicted experimental observations. For instance, the spectrum of a blackbody peaks at a finite wavelength and decreases at shorter wavelengths, as shown by experimental data.

Despite its limitations, the Rayleigh-Jeans law was a pivotal contribution to the study of blackbody radiation. Its shortcomings illuminated the limits of classical physics and prompted physicists to reconsider classical assumptions about energy distribution. This led Max Planck to propose a quantum theory of radiation in 1900. Planck introduced the idea that energy is quantized and can only be emitted or absorbed in discrete packets, or quanta, proportional to the frequency.

$$E = hf$$

Where h is Planck's constant (6.626×10^{-34} Js), and f is the frequency of radiation. In turn, Planck postulated that the oscillators in the blackbody, which are responsible for the radiation, can have discrete values only, given by $E_n = nhf$, where n = 0,1,2...

The probability of certain energy of oscillators depends on the Boltzmann distribution, $P(E_n) \propto e^{-E_n/k_BT}$, where k_B stands for Boltzmann's constant. Thus, the average oscillator energy $\langle E \rangle$, which is the sum over all weighted possible states or microstates of energy level occupation, is computed by substituting the relation for energy value E_n. This leads to

$$\langle E \rangle = \frac{\sum_{n=0}^{\infty} nhf \, e^{-nhf/k_BT}}{\sum_{n=0}^{\infty} n \, e^{-nhf/k_BT}}$$

Using the properties of geometric series simplifies this to: letting $x = e^{-nhf/k_BT}$, the denominator $\sum_{n=0}^{\infty} x^n$ evaluates to $\frac{1}{1-x}$, and the numerator $\sum_{n=0}^{\infty} nx^n$ evaluates to $\frac{1}{(1-x)^2}$. Substituting these into the expression gives $\langle E \rangle = \frac{hf}{e^{hf/k_BT}-1}$ the average energy of an oscillator.

The spectral energy density, $u(f,T)$, describes the energy emitted by a blackbody as a function of frequency and temperature. Planck's Law in terms of frequency is:

$$u(f,T) = \frac{8\pi f^2}{c^3} \cdot \frac{hf}{e^{\frac{hf}{k_BT}}-1},$$

where c is the speed of light. Alternatively, using $f = c/\lambda$, it can be expressed in terms of wavelength:

$$u(\lambda, T) = \frac{8\pi hc}{\lambda^5} \cdot \frac{1}{e^{\frac{hc}{\lambda k_B T}} - 1}$$

It revolutionized physics by introducing the concept of energy quantization, where energy is emitted or absorbed in discrete packets called "quanta." The amount of energy in each quantum is directly related to the frequency of the radiation, meaning higher frequencies carry higher energy. The law shows that the intensity of radiation emitted by a blackbody increases with temperature, and the peak of the emission shifts to shorter wavelengths, or higher frequencies, as the temperature rises. Planck's Law also demonstrated that classical physics, which predicted infinite energy at high frequencies (the "ultraviolet catastrophe"), failed to describe the observed behaviour of blackbody radiation.

The other factor which leads all these to birth of quantum mechanics was Einstein's explanation of the photoelectric effect. It was groundbreaking because it challenged the classical wave theory of light. Before Einstein, light was thought to behave purely as a wave, and this theory could not explain certain phenomena like the photoelectric effect. The photoelectric effect occurs when light strikes a metal surface and causes the emission of electrons. Classical wave theory predicted that increasing the intensity of light should eject electrons, regardless of the light's frequency. However, experiments showed that no electrons were ejected if the light's frequency was below a certain threshold, even if the light was very intense. This puzzled scientists and suggested that the wave theory of light was incomplete.

Einstein addressed this issue by proposing that light behaves not only as a wave but also as a particle, which he called a photon. According to his theory, light is composed of discrete packets of energy, each corresponding to a specific frequency. The energy of a photon is proportional to the frequency of the light, and this relationship is expressed by the equation $E = h\nu$, where E is the energy of the photon, h is Planck's constant, and ν is the frequency of the light. Einstein's key insight was that when light shines on a metal surface, it is the photons, not the wave nature of light, that interact with the electrons in the metal.

For an electron to be ejected from the metal, the photon must have enough energy to overcome the work function of the metal, which is the minimum energy required to free an electron. If the energy of the photon is greater than the work function, the photon transfers its energy to the electron, causing it to be ejected from the metal surface. If the photon's energy exceeds the work function by an amount ΔE, the excess energy is converted into the kinetic energy of the ejected electron. This explanation clarified why only light above a certain frequency, regardless of intensity, could eject electrons, as the frequency determined the energy of the photons.

Einstein's work on the photoelectric effect provided crucial evidence for the emerging theory of quantum mechanics, which posits that energy is quantized. It also helped establish the dual nature of light, which could exhibit both wave-like and particle-like properties depending on the situation. His explanation of the photoelectric effect earned him the Nobel Prize in Physics in 1921.

In 1923, Arthur Compton, an American physicist, made the discovery, which further defined the course of modern physics. He was investigating the interaction of X-rays with matter when he observed something that surprised him. He aimed X-rays at electrons and found that the X-rays were scattered at different angles but showed an important characteristic: their wavelength had increased, or their energy had decreased after collision. This was puzzling because, according to classical wave theory, it was not possible for light to be lost in such a manner by a simple collision.

It was only when Compton was working on this problem that he realized that the explanation lay in the fact that X-rays were behaving like particles. He proposed that the X-rays, which are now called photons, collided with the electrons and transferred some of their energy, causing the scattered X-rays to have longer wavelengths. By treating the interaction as a collision of a particle, similar to how billiard balls transfer energy when they collide, Compton demonstrated that both the conservation of energy and momentum could explain the phenomenon. His work provided solid evidence for the particle theory of light, adding a new layer of complexity to our understanding of the nature of light.

This discovery is known as the Compton effect. Despite its successes, the old quantum mechanics, which included the work of Planck and Bohr, had limitations that were becoming increasingly apparent. It could explain certain phenomena like the spectral lines of hydrogen atoms or the quantization of energy levels, but it failed to address more complex systems, particularly when dealing with multi-electron atoms or the behaviour of matter on a microscopic scale. The old quantum mechanics relied heavily on classical principles and the assumption that particles followed well-defined paths. However, this approach could not account for phenomena such as the wave-particle duality of matter or the Heisenberg uncertainty principle, which states that we cannot simultaneously know both the position and momentum of a particle with absolute certainty. These fundamental gaps pointed toward the need for a more comprehensive theory.

In the quantum mechanical framework, energy is not viewed as a continuous variable but rather as quantized. This means that energy exists in discrete packets or levels, and particles such as electrons can only occupy certain allowed energy states. This quantization explains a variety of phenomena, such as the discrete energy levels observed in atoms. The quantum state of an electron, for example, is defined by its energy level and its probability distribution around the nucleus, not a definite trajectory as classical mechanics would suggest. The uncertainty inherent in these quantum states, combined with the idea that particles also exhibit wave-like behaviour, led to a radical shift in our understanding of the microscopic world.

One of the most significant drawbacks of old quantum mechanics was its reliance on classical concepts, which were incompatible with emerging experimental evidence. The Bohr model, for instance, successfully explained the hydrogen atom's spectral lines but failed to account for more complex atoms, especially those with multiple electrons. Moreover, the concept of orbits in which electrons exist was inconsistent with the wave nature of particles. Classical physics also couldn't explain the behaviour of electrons in atoms beyond the simplest hydrogen atom or the nature of chemical bonding. As a result, the old quantum mechanics was eventually superseded by a more complete theory—quantum mechanics—developed by pioneers like Werner Heisenberg, Erwin Schrödinger, and Paul Dirac.

The dual nature of light was another cornerstone of quantum mechanics. Light, as shown by the photoelectric effect and Compton scattering, exhibits both wave-like and particle-like properties. This duality is encapsulated in the idea that light and other quantum particles, like electrons, do not fit neatly into classical categories of "wave" or "particle." Instead, they are described by wavefunctions, which provide a probability distribution for where a particle might be found or what its momentum might be. This wave-particle duality was an essential concept in the formation of quantum mechanics, where particles such as electrons are not viewed as point-like objects but as entities with wave-like characteristics that can be described by a wavefunction. The culmination of these developments led to the formation of quantum mechanics, a theory that describes the behaviour of matter and energy on the atomic and subatomic levels.

2: ARE WE WAVES OR PARTICLES?

What are we really made of? The answer would seem simple at first glance: flesh, bones, and a mind that does think. But peel back the layers, dive deep into the quantum realm, and a far stranger picture seems to emerge. Everything, from the tiniest electron to the largest molecule, seems to exist equally as a particle and a wave. This duality isn't just some kind of quirk of physics; it's a doorway to a deeper understanding of existence itself.

Let me illustrate it with a classical analogy: Consider tossing a coin. Before tossing, we would say that the probability of falling on its head or its tail is about 50/50. After the coin lands, one can easily observe whether it has fallen on its head or its tail. But right now, as this coin is in the air, spinning, what side of the coin is up? It is both in a way, depending on how the mind views it. This, of course, is only a simple, imperfect analogy aimed at helping the mind try to visualize duality. Now, for particles, this gets far more interesting and intricate. Particles have a very strange and puzzling feature: duality. Some scientists firmly believed that

BEYOND THE ATOMIC LEVEL

the particles are point-like solid objects, while others insisted they were waves, spreading out in space. And then, a new perspective arose: what if particles are both of these?

Imagine standing in a dimly lit room with two narrow slits cut into a barrier. Beyond the barrier lies a screen that can record impacts. Now, let's start by imagining what happens if we use particles, like tiny marbles.

Throw a handful of marbles through the slits at random. Some will hit the barrier, while others will pass through the slits and form two distinct bands on the screen, one behind each slit. Simple enough, right? Now, let's replace the marbles with light.

Well, when we pass light through these slits, something very interesting happens.

Light is a wave, and waves passing through each slit spread out and interfere with each other. This interference creates an interference pattern on the screen—a series of bright and dark fringes. Bright fringes are caused by constructive interference, in which the light waves reinforce each other, and dark fringes are caused by destructive interference, in which the light waves cancel each other out. That would be what we call an interference pattern and has been well understood since the days of Thomas Young's experiments in the early 19th century.

Then they did the same experiment but with electrons. Of course, an electron is a particle and could behave like marbles. We should get two bands on the screen as we got from marbles. And, to start off with, that expectation is met. If we fire one electron, it would fall somewhere on the screen. But as more electrons are fired, a pattern begins to emerge—not two distinct bands, but the same interference pattern we saw with light waves!

It's as though the electrons are behaving like waves, passing through both slits simultaneously, interfering with themselves, and creating the interference pattern. This is the first crack in the intuitive understanding of nature. How can something with mass act like a wave?

Scientists decided to check further. They put detectors at the slits to determine which slit the electrons went through. After all, if electrons are particles, they must go through one slit or the other, right?

The moment they did this, the interference pattern disappeared. Instead of acting like waves, the electrons began behaving like particles again, forming two bands on the screen. In other words, the act of observing or measuring the electrons changed their behaviour.

Let that sink in for a moment. The act of observation—only the mere attempt to see which slit it passes through—altered the outcome. This phenomenon is called the observer effect, and it lies at the very center of quantum physics.

Are electrons waves? Are they particles? Is it both? It depends on how one observes the electrons. Unobserved, electrons exist in a kind of probabilistic wave-like state of superposition, and they can go through both slits at once; but when observed, the superposition collapses into a single, definite state—a particle through one slit or the other. This experiment brought into light the fundamental duality of nature: particles can act as waves, and waves can act as particles.

It is here that Louis de Broglie stepped in. He came up with a revolutionary theory that combined these two apparently contradictory views into one—proposing, in fact, that particles can indeed act as waves under certain conditions. In 1924, a young physicist named Louis de Broglie came up with a proposition so seemingly absurd at the time: particles, like electrons, could behave like waves.

Up to that time, the wave-particle duality of light was well known—light sometimes acted like a wave and at other times like a stream of particles called photons. But de Broglie extended this concept even to all matter. It was his contention that it was a property of every particle independent of size, having a related wavelength given by the following equation:

$$\lambda = \frac{h}{p}$$

At the time, this was a wild and crazy hypothesis: Could particles like electrons, which we think of as tiny billiard balls, really have wave-like properties? This one question opened the door to a new reality—a reality that challenges the very way we think about the world.

What kind of waves are these? Do they resemble water waves, rippling through space? Not quite. De Broglie waves are fundamentally different. They are not physical waves but mathematical constructs, representing the probability of finding a particle in a certain location. An electron, for instance, when not observed, isn't in one specific place but is distributed as a "cloud of possibilities." When measured, this cloud collapses into a single point, revealing the particle's exact location.

This idea is deeply unsettling. It suggests that particles—and by extension, everything in the universe—are not "real" in the way we typically imagine. Until they are observed, their existence remains spread out, uncertain, and fundamentally wave-like.

Now, let's pause for a moment. If even the most basic building blocks of reality don't exist in a definite state until they're observed, might this also apply to us? Could our very existence depend on how we interact with the world?

To understand this better, let's explore the mathematics. De Broglie's relation isn't just an abstract idea—it naturally arises from established principles of physics.

Einstein's famous equation for a photon is $E = pc$, where p is momentum and c is the speed of light. Planck's equation, $E = h\nu$, relates energy to frequency. Combining these, de Broglie reasoned that if light exhibits both wave and particle properties, then matter might as well. By equating $\nu = \dfrac{E}{h} = \dfrac{pc}{h}$, he arrived at the now-famous relation.

This elegant equation is far more than a mere formula—it represents a profound insight into the universe. How can such a simple relationship describe both the solidity of particles and the fluidity of waves? It's as if nature itself is written in the language of duality, urging us to question what is real and what is illusion.

But how does this duality manifest? If particles are waves, why don't they spread out everywhere? The answer lies in the concept of wave packets.

A wave packet is a packet of waves localized in space, created by the superposition of many plane waves with nearly equal wavelengths. This combination of waves produces a pattern where the waves reinforce each other in one small region and cancel each other elsewhere, resulting in something that appears like a particle localized in space while maintaining its wave-like nature.

This raises an intriguing question: if particles aren't confined to one specific point until observed, could this principle apply to us as well? Are we, too, a superposition of possibilities, only defined by our actions and interactions with the world?

De Broglie waves introduce yet another layer of complexity through their velocities. These waves can be described by two kinds of velocity:

Phase velocity (v_p), which describes how the crests of individual waves move.

Group velocity (v_g), which represents the speed at which the wave packet—and therefore the particle itself—travels.

For de Broglie waves, the phase velocity is strange—it can exceed the speed of light! But before getting too excited, this doesn't violate relativity because the phase velocity doesn't carry information. The group velocity, which represents the particle's actual motion, always respects the speed of light limit.

Still, it's worth pondering: if parts of a wave-particle system can, in some sense, exceed the speed of light, does this hint at a reality operating beyond our current understanding?

Evidence supports this idea. In 1927, Clinton Davisson and Lester Germer provided the first experimental confirmation of de Broglie's hypothesis. They fired electrons at a nickel crystal, expecting random scattering. Instead, the electrons created a diffraction pattern—a hallmark of wave behaviour. This was undeniable proof that electrons, once thought to be solid particles, also behave like waves.

This finding didn't just validate de Broglie's theory—it shattered the classical view of matter. If electrons could act like waves, what about larger particles? Could wave-particle duality apply universally, even to us?

As experiments advanced, researchers discovered that even larger particles, like atoms and molecules, could exhibit diffraction patterns under the right conditions. This suggested that wave-particle duality wasn't confined to the microscopic but applied universally.

Here's the big question: if larger objects can behave like waves, where do we draw the line? At what point do we stop being "particles" and start being "waves"? Or is this distinction itself an illusion?

This exploration of wave-particle duality challenges our understanding of reality. It suggests a universe not solid or deterministic but one shaped by probabilities and possibilities, deeply influenced by observation and interaction.

This raises profound philosophical dilemmas. What is reality if it cannot be pinned down with certainty? If particles exist in multiple states simultaneously, how do we define "what is"? Are we in a universe that becomes definite only when observed, or is reality fixed, awaiting our full perception?

Philosophers have long pondered duality, not just in physics but in human nature. Ancient Greek philosopher Plato, for example, saw humans as divided between the rational soul (logos) and the irrational soul (pathos). The rational governs logic and reason, while the irrational encompasses emotions and instincts. For Plato, a harmonious balance of these forces defined the ideal human life, though tension between them often creates internal struggles.

This philosophical duality parallels the scientific idea of wave-particle duality, suggesting that both humans and the universe may exist in a state of fundamental duality, defined by their interactions and perceptions.

This duality remains deeply relevant to modern life. In our daily decision-making, we often find ourselves torn between what we should do (the rational, logical approach) and what we want to do (the emotional, impulsive side). Philosophers like Immanuel Kant and René Descartes also wrestled with this divide. Descartes famously declared, "Cogito, ergo sum" (I think, therefore I am), emphasizing that human existence is rooted in rational thought—possibly at the expense of acknowledging the emotional and subjective aspects of being.

This tension has fueled debates and philosophical discussions that intersect with religious and spiritual beliefs, including arguments about the existence and role of God. Many monotheistic traditions like Christianity, Islam, and Judaism emphasize the idea of a single, divine truth governing the universe. The indeterminacy suggested by quantum mechanics—probabilities, superpositions, and multiple possibilities—challenges the deterministic belief that God has a clear, preordained plan for everything.

In religious contexts, God is often seen as the ultimate

source of order and purpose, with a divine plan guiding the universe. Yet, quantum mechanics introduces uncertainty: particles existing in multiple states at once (superposition) and outcomes influenced by observation (the observer effect). This raises profound questions: If the universe is inherently probabilistic, does it challenge the idea of a divine plan? Is God merely an observer, or could the universe reflect a form of divine chaos?

Some argue that the unpredictability of quantum mechanics conflicts with the notion of a perfectly ordered creation. Albert Einstein, for example, famously remarked, "God does not play dice with the universe," expressing discomfort with the idea of a universe governed by chance rather than predictability.

Quantum mechanics' duality also adds complexity to the concept of human agency. If our universe operates on uncertainty until observed or measured, does this suggest we have true free will? Or are our choices and actions part of a broader divine plan?

For some, this uncertainty is unsettling, as it implies a universe that might be chaotic and unpredictable, undermining the notion of clear-cut divine control. Others argue that free will and uncertainty coexist as part of a greater design. Perhaps God allows indeterminacy to preserve human autonomy. In this view, uncertainty doesn't diminish God's role but reflects a dynamic interplay between divine will and human choice.

So, what are we? Deterministic particles, bound by cause and effect? Or waves, fluid and probabilistic, shaped by our choices and interactions? Perhaps, like the particles we study, we embody both—a reflection of the duality woven into the fabric of the universe itself.

3: THE GHOST IN THE ATOM

"Anyone who is not shocked by quantum theory has not understood it."
— *Niels Bohr*

The atom is the constituent unit of everything that exists around us, from the universe downwards. Life can get overwhelming sometimes, but trying to put our issues into perspective with the enormity of the universe out there will surely help. As a matter of fact, as compared to the universe or even large celestial bodies, our problems become immensely small.

People are more often fascinated by big things: big houses, expensive cars, or large sums of money. Yet, rarely do we think about the smallest elements of existence; atoms are one of these tiny things. If Earth were a huge playground, then an atom would be just like a small moving football in it.

Just as a change in one's view of life can make the problems seem much smaller, taking into consideration the size of an atom is a humbling thought. However, unlike our problems, the characteristics of an atom do not change with our perception of them; they remain constant-acting, often in a manner quite contrary to what we might expect.

Many scientists have delved deep into the atomic level to unravel the mysteries of this tiny yet fundamental entity.

BEYOND THE ATOMIC LEVEL

From Acharya Kanad, who first proposed the existence of indivisible particles, to Schrödinger, who introduced wave mechanics, the journey of understanding the atom has been replete with path-breaking theories that were later challenged or refined by further discoveries.

Classical mechanics initially tried to explain the atom in simple, deterministic terms, but its inadequacy became increasingly evident as the subtleties of atomic behaviour were revealed. This opened the way to quantum mechanics, a field that redefined our understanding of nature at the smallest scales, often in a manner quite unintuitive and in conflict with classical expectations.

The classical atom, represented by such models as Thomson's "plum pudding" and the nuclear model of Rutherford, was an attempt to describe the atom in terms of deterministic laws of motion. This treated electrons as particles orbiting the nucleus, somewhat as planets do around the Sun, at the mercy of predictable, classical mechanics. Such was the classical view of matter that could not explain fully the facts in the explanation of spectral lines or the stability of atoms.

Quantum mechanics revolutionized these views by introducing wave-particle duality of electrons, besides making atomic behaviour probabilistic. Schrödinger's wave equation replaced the well-defined orbits of the classical models with "orbitals," regions in which the electrons have the probability of being found. Quantum mechanics avoids certainty for an atom and its dealings in a mathematical manner but gives an accurate view related to atomic structure and dealing.

The transition from Classical to Quantum Physics marked a revolution in the understanding of the universe. In contrast, Classical Physics is based on determinism; if the initial conditions of any system are ascertained, its future behaviour can be predicted. The advent of quantum mechanics saw the inculcation of probabilistic interpretations in lieu of this determinism.

Heisenberg's uncertainty principle perhaps epitomized this change by stating that one cannot know both the position and momentum of a particle at any instant of time with perfect accuracy. Inherent uncertainty thus blew up the deterministic view and changed it with the reality of probability wherein the very act of measurement influences the outcome.

In 1913, Niels Bohr introduced a revolutionary model of the atom, expanding on Rutherford's earlier nuclear structure. Bohr's model proposed that electrons occupy quantized energy levels, successfully explaining phenomena like atomic spectra.

According to this model, electrons orbit the nucleus in specific circular paths, or energy levels. These levels are quantized, meaning only certain discrete orbits are possible, each associated with a specific energy. The atom remains stable as long as an electron stays within one of these fixed orbits, without radiating energy—contradicting classical electrodynamics predictions.

A key aspect of Bohr's model is that electrons can move between these energy levels by absorbing or releasing energy. When an electron absorbs energy, it jumps to a higher level, and when it emits energy, it drops to a lower one. This explained the discrete spectral lines observed in hydrogen's emission spectrum.

Bohr also introduced the idea that an electron's angular momentum in a stationary orbit is quantized. Specifically, the angular momentum is an integer multiple of $h/2\pi$, where h is Planck's constant. This quantization was essential in determining the size and energy of allowed orbits, paving the way for modern quantum theories.

Before Bohr, classical physics failed to account for the distinct lines in the hydrogen spectrum, predicting a continuous spectrum instead. Bohr's quantized energy model addressed this limitation, laying the foundation for the quantum mechanics revolution.

Before Niels Bohr's work, scientists struggled to explain the discrete spectral lines observed in the hydrogen spectrum, as classical physics predicted a continuous spectrum of radiation. Bohr's model transformed this understanding by introducing quantized energy levels for electrons, successfully explaining the spectral lines and laying the groundwork for modern quantum theory. In Bohr's model, electrons occupy fixed, quantized energy levels around the nucleus, with each level corresponding to a specific circular orbit. The energy of each orbit is determined by the formula
$$E_n = -\frac{13.6 eV}{n^2},$$
where E_n is the energy of the nth orbit, 13.6 eV represents the ground state energy of the hydrogen atom, and n is the principal quantum number (n=1,2,3,...). The lowest energy level (n=1) is called the ground state, while higher levels (n>1) are excited states. Electrons can transition between these energy levels by absorbing or emitting a photon whose energy equals the difference between the initial and final levels. The energy of the photon is described by
$$E_{photon} = \frac{13.6 eV}{n_i^2} - \frac{13.6 eV}{n_f^2},$$
where n_i and n_f are the initial and final principal quantum numbers. This quantized approach explains the discrete nature of hydrogen's spectral lines, as each emitted photon corresponds to a specific wavelength of light, producing distinct lines in the spectrum. By addressing the shortcomings of classical physics, Bohr's model provided a foundational understanding of atomic structure and emission spectra.

Bohr's model successfully explained the visible spectrum of hydrogen, known as the Balmer series. In this series, when

BEYOND THE ATOMIC LEVEL

an electron in a hydrogen atom transitions from a higher energy level (n>2) to n=2, it emits photons that form visible lines in the spectrum. The wavelength of these emitted photons is determined using the Rydberg formula:

$$\frac{1}{\lambda} = RH \left(\frac{1}{n_f^2} - \frac{1}{n_i^2}\right)$$

Here, λ is the wavelength of the emitted photon, RH is the Rydberg constant (RH=1.097×10^7 m^{-1}), n_f=2 is the final energy level, and n_i is the initial energy level (ni>2). This formula explains the wavelengths of light emitted as electrons transition to the second energy level. The visible lines in the hydrogen spectrum, such as red, blue, and violet, correspond to transitions from higher levels (n=3,4,...) to n=2.

Beyond the Balmer series, Bohr's model also accounted for other spectral series of hydrogen. The Lyman series describes transitions where electrons fall to n=1, producing ultraviolet light, while the Paschen series involves transitions to n=3, resulting in infrared light. These series are similarly described by the Rydberg formula, with n_f and n_i taking on values specific to the transitions in question. Bohr's explanation of these series was a critical validation of his model and a significant contribution to the understanding of atomic spectra.

Bohr's model, although revolutionary in its explanation of the hydrogen atom, had severe limitations when applied to multi-electron systems. The model's assumption of electrons moving independently in well-defined orbits was inadequate in the presence of electron-electron interactions, which are crucial in atoms containing more than one electron. In multi-electron systems, electrons experience Coulomb repulsion, and it is impossible to treat their motion independently. This interaction leads to phenomena, such as shielding, through which inner electrons reduce the effective nuclear charge experienced by outer electrons. Bohr's model couldn't address the fine structure of the atomic spectra caused by the relativistic effect and by spin-orbit coupling; this was because it hadn't been capable of successfully predicting the spectrum of atoms involving more than one electron in the atom, which really proved the flaws of Bohr's atomic model. Furthermore, Bohr's model treated electrons as particles in fixed orbits, a concept that failed to incorporate the wave-particle duality of electrons, which is central to quantum mechanics. Electrons exhibit both particle-like and wave-like properties, and their behaviour in multi-electron systems is better described by wavefunctions and orbitals. Bohr's model also did not cope well with increasing electron configuration complexity in the case of atoms with many electrons. Repulsive forces between the electrons change the energy levels and add even more complications. The electron energy level treatment of the model was insufficient to explain the complicated relationships in the multi-electron atoms, especially with increasing numbers of electrons. Quantum mechanics developed due to the inability to explain relativistic effects, spin-orbit coupling, and the wave nature of electrons. Quantum mechanics, in its probabilistic approach, with the use of quantum numbers, provided a better description of electron behaviour. The Schrödinger equation, and the Pauli exclusion principle provided an explanation to the arrangement of electrons within multi-electron systems resolving many of the issues not explained by the Bohr model. Thus, through quantum mechanics, the sophistication of atomic structure was approached and the simple orbits within Bohr's model replaced by more complex wavefunctions and energy levels in explaining the multi-electron systems.

After de Broglie's concept, wave-particle duality, atomic and subatomic particles' understanding became something different. He himself forwarded the proposition that, "if light could both particle-like and wave-like property, then it's pretty possible for the same reasons that electrons may share its properties". It started one new dimension in understanding how atoms work. However, though de Broglie's hypothesis marked the turning point, still it was evident that some overall framework to explain how particles behave, especially in the case of electrons inside atoms, was missing. Classical models, such as Bohr's atomic theory, could not explain wave-particle duality on a consistent basis. Bohr's model of the atom, although brilliant in explaining the hydrogen atomic spectral lines, did not account for the behaviour of more complex atoms. Now, electron-electron interaction, relativistic effects, and quantum mechanical principles will have to be accounted for. This led to the following philosophical dilemma: were the deterministic views of classical physics, which conceived the electrons as particles moving on well-defined orbits, indeed adequate to explain phenomena that are observed at the atomic scale? Or was a more profound understanding of physics still hidden?

The key question emerged: How could we combine the wave-like behaviour of particles with the quantized nature of atomic spectra?

This duality in nature was pointed out by de Broglie and signified that something more novel must be created for theoretical purpose. That is, not just some little amendment of existing theories, but the total rewriting of what was understood for atomic and subatomic level of matters. The greater they would explore in quantum regions, the more it came out to be an underestimation to their level. Classical models were based on deterministic principles where, if the initial conditions of a system were known, the future could be predicted with certainty. Yet, this predictability began to break down at the atomic scale, where the uncertainty of position and momentum became increasingly apparent.

Enter Erwin Schrödinger, whose insight was to provide the needed link between the classical and quantum worlds.

BEYOND THE ATOMIC LEVEL

Schrödinger's work, while being a development of de Broglie's wave-particle duality, was directed toward a more mathematically profound description of the electron's behaviour that could explain its wave-like character while retaining discrete energy levels, such as those of the hydrogen atom. His formulation required abandoning the classical view of electrons as particles moving along fixed paths and embracing a more complex, probabilistic interpretation of their behaviour.

The transition from a deterministic worldview, where every element of nature could be known with certainty, to a probabilistic worldview, where only the likelihood of an event could be determined, represented a philosophical revolution. This shift didn't just challenge physics; it challenged our understanding of reality itself. Schrödinger's wave mechanics, as represented by the wave function, provided an ideal of a world where the electron could not be located in space precisely at any point in time but its likelihood of finding it in any given position could be described. These implications were very far beyond physics, stretching into philosophical thought and into how people perceive the nature of knowledge and reality.

Schrödinger's wave mechanics was what opened the quantum world for understanding. His formulation allowed for a mathematical framework through which one could describe the behaviour of electrons in atoms, no longer as particles in definite orbits but as waves of probabilities. The new model, which involved wave functions and uncertainty, ushered in a new era of atomic theory. As we move from de Broglie's wave-particle duality to Schrödinger's wave mechanics, the very nature of the atom and the electron was transformed from a deterministic entity to a probabilistic wave, and with this new paradigm, Schrödinger unlocked the deeper mysteries of the atom.

Particles such as an electron are not like how we find in

school texts where they orbit. They are much easier to get hold of when they are huge bodies in space. The smallest things are the hardest things to get hold of. The more we go beneath, the harder it gets to catch something small. We can only predict their position through probability. The smaller that object is, the more uncertainty. To explain this, the Schrödinger derived an equation, namely the Schrödinger equation, which governs the time evolution of the wave function for a particle and, as such, one can predict the probability with which it is possible to find that particle, let's say an electron. Rather than thinking of electrons as particles moving in fixed orbits, Schrödinger's equation treats them as wave-like entities, where their position is described by a probability distribution. This approach helps understand the uncertainty and wave-particle duality of subatomic particles.

To understand Schrödinger's wave mechanics, we start by combining classical mechanics with quantum theory to describe the behaviour of particles like electrons. This description is framed in terms of a wave function ($\psi(x)$), which represents the probability amplitude of finding the particle at a specific position. The square of the wave function's magnitude ($|\psi(x)|^2$) gives the probability density at that location.

In classical mechanics, the total energy (E) of a system is the sum of its potential energy (V) and kinetic energy (T), expressed as $E=T+V$. For a particle of mass mm, the kinetic energy is given by $T = \dfrac{p^2}{2m}$, where p is the particle's momentum. Quantum mechanics relates this momentum to the wave nature of particles, with momentum $p=\hbar k$, where $\hbar$ is the reduced Planck's constant and k is the wave number of the associated wave. Substituting $p=\hbar k$ into the kinetic energy equation gives

$$T = \frac{\hbar^2 k^2}{2m}$$

BEYOND THE ATOMIC LEVEL

In quantum mechanics, a particle is not defined by precise position or momentum but by a wave function, $\psi(x)$, which encapsulates the particle's probabilistic nature. The particle's total energy in this framework remains the sum of kinetic and potential energies but is expressed in terms of the wave function. Mathematically, this is written as:

$$E = \frac{\hbar^2}{2m} \frac{\partial^2}{2x^2} \psi(x) + V(x)\psi(x)$$

Here:

- $\frac{\hbar^2}{2m} \frac{\partial^2}{2x^2} \psi(x)$ represents the kinetic energy, where $\frac{\partial^2}{2x^2} \psi(x)$ is the second derivative of the wave function, indicating its spatial curvature.
- $V(x)\psi(x)$ represents the potential energy of the particle at position x.

Reorganizing this equation leads to the time-independent Schrödinger equation:

$$H^\wedge \psi(x) = E\psi(x)$$

In this equation, $H^\wedge$, the Hamiltonian operator, encapsulates the total energy of the system, combining both kinetic and potential energy terms. The Hamiltonian operates on the wave function ($\psi(x)$) to yield the system's energy (E). This fundamental equation underpins quantum mechanics, describing how particles behave at microscopic scales and how their probabilistic nature is governed by wave functions.

To summarize conceptually, Schrödinger's equation provides a framework for describing particles not in terms of definite positions or momenta but through a wave function that gives the probabilities of finding them at various locations. It integrates quantum mechanics and wave theory, replacing the deterministic outlook of classical mechanics with a probabilistic interpretation of particle behaviour at atomic scales.

Schrödinger's equation is a cornerstone of quantum mechanics, explaining how the wave function of a quantum system evolves. It offers insights into the probable positions and energies of microscopic particles, such as electrons in atoms. The equation has two primary forms: the time-dependent Schrödinger equation and the time-independent Schrödinger equation. The time-dependent version is the most general, describing systems where properties change over time or where potential energy varies with time. The time-independent version is suited to analyzing systems in stationary states where the energy levels are constant.

The time-dependent Schrödinger equation is expressed as:

$$i\hbar \frac{\partial \psi(x,t)}{\partial t} = H^\wedge \psi(x,t)$$

Here:
- i is the imaginary unit ($i^2 = -1$),
- $\hbar$ is the reduced Planck's constant ($h/2\pi$),
- $\frac{\partial \psi(x,t)}{\partial t}$ is the time derivative of the wave function, indicating how it changes over time,
- $H^\wedge$ is the Hamiltonian operator, representing the system's total energy (kinetic + potential),
- $\psi(x,t)$ is the wave function, a complex function encoding the probability amplitude of finding the particle at position x and time t.

This equation comprehensively describes a quantum system in motion. The wave function $\psi(x,t)$ encapsulates all information about the particle's state, including where it is likely to be found. The equation shows that the rate of change of the wave function over time depends on the system's Hamiltonian. By solving this equation, we can predict how the probability distribution of a particle evolves, making it an essential tool in quantum mechanics for understanding dynamic systems.

The time-dependent Schrödinger equation is the very cornerstone of investigations dealing with time-dependent phenomena within quantum mechanics. For example, such a

BEYOND THE ATOMIC LEVEL

description can model the dynamics of electrons in atoms under the action of an electric or magnetic field changing in time. In that case, the potential energy V(x,t)V(x,t) depends on time, and the wave function evolves accordingly. The given equation also plays an important role in those systems where, due to external forces or interactions, the properties of the system change with time.

While the time-dependent Schrödinger equation is completely general and applies, in principle, to any quantum system, it is often useful to restrict attention to the time-independent Schrödinger equation for a particle whose potential energy does not depend explicitly on time. This form is of especial value when considering states that are stationary in the sense that their properties do not change with time. Such stationary states are solutions to the so-called time-independent Schrödinger equation:

$$H^\wedge \psi(x) = E\psi(x)$$

The time-independent Schrödinger equation reduces the time-dependent equation on an assumption that the potential energy does not change with time. This allows us to ignore the time dependence of the wave function and consider only its spatial dependence. For these stationary states, the probability distribution of the particle at a point in space does not depend on time, whereas the wave function has a different phase at different times, irrelevant for the probability density.

The time-independent Schrödinger equation is particularly very useful in situations where one is interested in finding out the allowed energy levels for a quantum system and the corresponding wave functions, or eigenfunctions. For instance, an electron in an atom cannot occupy any energy level as it can only occupy those discrete energy levels. The solution to the time-independent Schrödinger equation for the electron in an atom gives us these quantized energy levels and their corresponding wave functions, which describe the probability distribution of the position of the electron in space.

The time-independent Schrodinger equation can also determine the allowed energy states in systems such as particles inside a box, harmonic oscillators, or quantum wells, depending on the nature of a constant or spatially-position-dependent potential, but independently of time. In both these cases, solving it gives us energy values determined as discrete, while the corresponding solutions to the wave functions describe stationary states where its properties do not change at all with time.

In the actual scenario, time-dependent and time-independent Schrödinger equations are connected but not separate entities. Time-independent Schrödinger equation can be derived under certain conditions from the time-dependent one. For most quantum systems, especially when dealing with bound states, like the case of an electron within an atom, potential energy is time independent; therefore, it is possible to split the time-dependent and spatial parts of the wave function.

When we try to solve a time-dependent Schrödinger equation, for many—in fact, for most practical cases—a separation of variables can be achieved. That is to say, it is frequently possible to construct a wave function which can be written as a simple product of a space part and a time-dependent part. Accompanying this separation, we find an independent time-less Schrödinger equation for the space-dependent part, plus a time dependence which does no more than introduce a phase factor $e^{-iEt/\hbar}$, where E is the energy of the system.

Hence, the time-independent Schrödinger equation is a special version of the time-dependent version when the system is under steady state. It represents a reduced version that would enable one to determine the eigenvalues of the energy of the system as well as its stationary wave functions.

BEYOND THE ATOMIC LEVEL

Both forms of the Schrödinger equation are necessary in quantum mechanics. The time-dependent Schrödinger equation is vital for studying dynamic systems, such as particles in external fields or evolving systems. On the other hand, the time-independent Schrödinger equation is often used in problems involving stationary states, where the system's potential does not change with time and the energy levels are quantized.

For example, the time-independent Schrödinger equation is heavily utilized in atomic physics for describing the behaviour of electrons within atoms, confined to specific energy levels. Similarly, it is applied in molecular and solid-state physics for describing properties of molecules and materials.

The time-dependent Schrödinger equation gives an account of how the wave function of a quantum system evolves in time, but the time-independent Schrödinger equation simplifies it to stationary states where all energy levels are fixed. Both are necessary for quantum behaviour at different scales or conditions.

We were talking about many things, and the common term we used to address these concepts was "wave function." Some people reading this might not fully understand what it is or might be confused. So, let's break it down.

Imagine a butterfly inside a closed box. We can hear the fluttering of its wings and, although we cannot see the butterfly, can guess where it might be from the sound. The waves of sound represent the uncertainty of where the butterfly is at any given time. Likewise, in quantum mechanics, the electron does not have a definite position as one would expect for larger objects. Instead, the electron position is represented in terms of probabilities using a wave function. While what we hear from the butterfly is due to the butterfly's motion, what the wave represents is the probability that might be seen by locating the electron at any place and time. The smaller the object (such as an electron), the more uncertain and unpredictable its behaviour becomes. We describe the position with probabilities rather than certainty because of this.

The wave function, ψ, in quantum mechanics, provides a description of a quantum system. For example, the state of one particle, which might be an electron, gives all of the information that is necessary in order to determine the probabilities for finding the particle at a given location in space and to get the characteristic energy value. This wave function contains a complex number function where the real and imaginary value parts may be separated.

While the wave function is not directly observable, its square magnitude, given by $|\psi(x,t)|^2$, provides the probability density of finding a particle at a particular point at time t.

For instance, if we are interested in the probability of finding a particle in the space between x1 and x2, we can integrate the square of the wave function over that range:

$$P(x1 \leq x \leq x2) = \int_{x_1}^{x_2} |\psi(x,t)|^2 \, dx$$

This formula means that the probability of finding the particle within the interval [x1,x2] is proportional to the integral of the square of the wave function over that range. In this way, the wave function provides all the information we need to predict the behaviour of quantum systems, even if we can't directly observe the particle's position. Instead, we use probabilities to describe where it might be at any given moment.

What does this say to us as conscious observers? If the universe works on probabilities, how does that play out in the world of our everyday experiences, where things seem to behave according to clear causes and effects? The idea that reality, at its most fundamental level, is probabilistic suggests a world that is more mysterious and less predictable than we might have thought. It challenges our need for certainty and control, making us face the limitations of human knowledge and perception.

In the classical Bohr model, electrons have been thought of as existing in fixed orbits around a nucleus, similar to planetary orbits around the Sun. Quantum mechanics, however has replaced this with the use of electron clouds and orbitals. Rather than going through definite paths, electrons must be described in terms of regions of space around a nucleus, known as orbitals, where only their positions and momenta can have definite values. The term "electron cloud" refers to the distribution probability of where an electron should be at any given moment. An orbital is a three-dimensional region where there is a high probability of locating an electron. Unlike the well-defined orbits in Bohr's model, orbitals do not have a set path. Instead, they are characterized by quantum numbers that define the energy, shape, and orientation of the cloud. For example, s, p, d, and f orbitals represent different shapes and are further described by their angular momentum quantum number.

This transition from fixed orbits to probabilistic orbitals represents a deeper understanding of quantum behaviour where the certainty of an electron's position is replaced by probability. The concept of orbitals and electron clouds is more in tune with experimental observations, like atomic spectra, and is required to explain chemical bonding and atomic structure in a more accurate and complex way than that in the Bohr model.

It has gone to the very philosophical depths of science and questions whether an atom is a real thing or just an abstraction. At first blush, atoms appear to be nothing if not real. They make up the material around us-from the furniture we touch to the food we eat-and their absence is unthinkable in the physical world. However, when we look at the language of modern physics, atoms are described not in terms of solid, tangible particles but as probabilities, wave functions, and mathematical models. This distinction raises profound questions about the nature of reality and the role of human perception in shaping our understanding of the universe.

The atom was then viewed in classical physics as an actual, indivisible chunk of matter with the emergence of atomic theory in the 19th century. The early work of John Dalton treated atoms as small, hard spheres that combined in simple ratios to form compounds. Though useful, it remained an approximation of reality. The quantum mechanical development led our understanding of the atom to a drastically different scenario. Instead of being seen as a tiny, solid object, the atom was revealed to be a complex, dynamic system of subatomic particles—electrons, protons, and neutrons—interacting in ways that defy our everyday experiences of matter.

The atom, according to quantum mechanics, is not an observable static object. Instead, it is described by wave functions that give a probabilistic description of where the electron is likely to be found. The electron is no longer imagined to be a small particle whizzing around the nucleus in a well-defined orbit but rather spread out in a "cloud" of probabilities, existing in a range of possible states until measured. This probabilistic nature, as encapsulated by Heisenberg's uncertainty principle, defies the deterministic, mechanistic view of the world that classical physics once promoted. It seems even more basic that we cannot know, at the same time, the exact position and the momentum of an electron, so that at the quantum level, reality is no longer a simple set of objects and events but is, rather, a web of probabilities, uncertainties, and interactions.

This raises the question: does the atom exist as a physical entity independent of observation, or is it merely a mathematical construct that represents our best understanding of the physical world? From a philosophical standpoint, the answer may depend on one's interpretation of quantum mechanics. One of the most widely accepted is the Copenhagen interpretation, which postulates that the wave function—the mathematical expression describing the state of the atom—does not correspond to a real, physical object but to a set of probabilities. Here, the atom's "reality" is only defined when it is observed or measured. Before it is measured, the electron is in a superposition of states and only collapses into a definite state upon observation. This then suggests that, at least in some sense, the atom is not "real" in the classical sense but rather a product of our interaction with the world.

Other interpretations, such as the many-worlds theory, present an alternative vision. This perspective is based on the assumption that all possible outcomes of a quantum measurement exist in parallel universes; in other words, the wave function represents a real but multiversal feature of reality. In this view, the atom remains a mathematical construct, but its potentiality is spread out in an infinite number of parallel worlds, each representing a different possibility for the atom's state.

Philosophically, these interpretations bring out the tension between the objective nature of the atom and the subjective nature of observation. If the properties of an atom are defined only by observation, then one would say that atoms, as all quantum objects, are merely mental constructs within our minds. This position comes under a form of idealism, where reality is a product of the perceptions of the mind. Conversely, the realist view is that the atom exists independently of human observation, and our imperfect mathematical models provide a window into a deeper, objective reality.

In this context, an atom is a symbol of the much wider philosophical debate between realism and anti-realism, beyond physics. Is reality a phenomenon that exists independently, or is it merely constructed in the mind based upon the observations and measurements of humans? The atom, both as physical object and mathematical model, poses these questions to us in deep ways. The more we know about the atom, the more we understand that reality is always filtered through the tools and models by which we describe it.

This tension is not specific to the atom. Indeed, the atom stands as a microcosm of the larger philosophical problem that all of science must confront: how to cross the gap between the abstract, mathematical descriptions of reality and the concrete, material world that we know directly. Whether the atom is "real" or merely a mathematical construct is, finally, a question that belongs at the intersection of science, philosophy, and metaphysics, and whose answer may depend on how one views the relationship between theory and reality itself.

The atom is one of the most singular things in the history of science and philosophy. It is a concrete thing that shapes the material world and a mathematical abstraction that challenges our understanding of the very nature of reality. Whether the atom is "real" or merely a construction is perhaps less important than the realization that our comprehension of the universe is always incomplete and mediated by our cognitive and observational limitations. The atom, in all its complexity, reminds us that the boundary between the real and the abstract is often more porous and fluid than we might like to admit.

BEYOND THE ATOMIC LEVEL

4: THE LIMITS OF KNOWING

All human activity began from a simple stone— breaking it, sharpening it, hunting animals, discovery of fire, language and money, and many other things. These happened just in a very short timeframe compared to the gigantic view of our universe. Even the universe is still young, compared to a child that can expand its imagination to beyond the unknown. Our universe is basically an overthinking child, which is also why it is hard to understand. It is always difficult to understand a child; their behaviour is unpredictable and sometimes out of our reach.

Humans are the most intelligent creatures on this earth, and we have always been striving to know more. We are fueled by an insatiable desire to learn. Does this pursuit of knowledge ever come to an end? We have often been selfish in our quest to understand, but as we learn more, we realize how little we truly know about reality. We cannot even determine the exact position of a tiny particle.

There, one of the most stunning and unsettling principles in all of modern physics comes alive: Werner Heisenberg's uncertainty principle. That is, according to classic thought, we believed with enough information, we'd be able to know absolutely the state of any system or part of a system at some point in time. With quantum mechanics, that assumption dissolves. Heisenberg's uncertainty principle tells us that there's a fundamental limit to exactly how precisely we can ever measure certain properties of these particles, such as both their position and momentum. The more accurately we can try to measure one, the less accurately we'll be able to measure the other; this is due to what's called inherent uncertainty within nature itself, not limitations in measurement tools. In this regard, the universe shows to be much more mysterious and unpredictable than one could have ever thought of.

It is as if the universe, like the overthinking child, is intentionally resisting our attempts to fully grasp its nature. We may never be able to fully understand the tiniest particles that make up the world around us, and perhaps the very act of observation itself influences the outcome. It is as if our will to know everything, our drive to reduce everything to certainty, constantly collides with the limits of our ability—and perhaps even with the limits of reality itself. This is the paradox of the uncertainty principle: the more we know, the more we realize how much is unknowable.

The more precisely we can determine one property, the less precise is our measurement of the other. This is not a consequence of poor instrumental design; rather, it is a direct result of wave-particle duality. Werner Heisenberg originally stated what has become known as the uncertainty principle: it is impossible to simultaneously measure with unlimited precision pairs of physical properties, such as position and momentum, that are connected to each other via the wave-particle dual nature of matter. That is, the more precisely we want to determine the position of a particle, the more indeterminate its momentum will be, and vice versa. Now, let's do some mathematics to understand it mathematically.

The expectation value of position is given by:

$$\langle x \rangle = \int_{-\infty}^{\infty} x \, |\psi(x)|^2 \, dx$$

Similarly, the expectation value of momentum ⟨p⟩ is given by:

$$\langle p \rangle = \int_{-\infty}^{\infty} \psi^*(x)\left(-i\hbar \frac{d}{dx}\right)\psi(x)\,dx$$

where $\hbar$ is the reduced Planck constant, and $\psi^*(x)$ denotes the complex conjugate of the wavefunction.

The uncertainty in position, Δx, is defined as the square root of the variance of x:

$$(\Delta x)2 = \langle x2 \rangle - \langle x \rangle 2$$

Analogously, the uncertainty in momentum, Δp, is the square root of the variance of momentum:

$$(\Delta p)2 = \langle p2 \rangle - \langle p \rangle 2$$

Now, to obtain the uncertainty relation, we have to express these uncertainties in such a form that relates to position and momentum.

The wavefunction in momentum space, $\psi \sim (p)$, is related to the wavefunction in position space $\psi(x)$ via the Fourier transform:

$$\psi \sim (p) = \frac{1}{\sqrt{2\pi\hbar}} \int_{-\infty}^{\infty} \psi(x) e^{-ipx/\hbar} dx$$

The uncertainty in momentum Δp is related to the spread of $\psi \sim (p)$ in momentum space, and similarly, the uncertainty in position Δx is related to the spread of $\psi(x)$ in position space.

The key mathematical relationship between position and momentum uncertainties follows from the Cauchy-Schwarz inequality. This inequality states that for any two functions f(x) and g(x), we have:

$$\left(\int_{-\infty}^{\infty} f(x)g(x)dx\right)^2 \leq \left(\int_{-\infty}^{\infty} |f(x)|^2 \, dx\right)\left(\int_{-\infty}^{\infty} |g(x)|^2 \, dx\right)$$

Applied to quantum mechanics, this inequality, taken for position and momentum operators, yields, after some manipulation and after an appeal to the Fourier transform, the Heisenberg uncertainty relation:

$$\Delta x \cdot \Delta p \geq \frac{\hbar}{2}$$

This uncertainty relationship states that the product of the uncertainties in position Δx and in momentum Δp is greater than or equal to $\hbar/2$. The reduced Planck constant sets the lower bound on the uncertainty.

This mathematical derivation demonstrates the fundamental link between uncertainties in position and momentum based on properties of the wavefunction and Fourier transform. The result of the derivation is the Heisenberg Uncertainty Principle, which says that it is impossible to precisely measure any one of these quantities while the other remains uncertain in some respect, no matter how precisely the first has been measured.

The principle reflects the wave-like nature of particles in quantum mechanics, where an attempt to localize a particle to a definite position increases the spread of its momentum, and vice versa. The principle has profound implications on our understanding of the quantum world, emphasizing that there is a fundamental limit to what can be known about a particle's state.

The uncertainty principle of quantum mechanics forces us to face the limits of human knowledge and understanding. For so long, the belief that the universe could be fully known by observation and measurement had been at the center of our worldview. Heisenberg's principle, however, tells us

BEYOND THE ATOMIC LEVEL

something very profound: there are some things that will always be unknowable. Regardless of how advanced our tools or theories might be, at the very core of nature lies fundamental uncertainty. It challenges not only our scientific approach but also our philosophical notions of knowledge itself: Can we ever really know anything completely, or is there always a shadow of uncertainty that defines our existence? Observations were no longer seen as a mere gathering of facts, but seemed to play an active role in shaping reality.

In quantum mechanics, the mere act of measurement seems to influence the state of the system, suggesting that reality is not an objective, fixed entity but something that is inextricably linked to the observer. This raises the question of whether reality exists independently of our perception of it, or if our understanding of the universe is forever limited by the constraints of human observation. In this light, the uncertainty principle forces us to reevaluate the nature of reality itself—an ever-changing, elusive phenomenon that may forever be beyond our grasp. In quantum mechanics, the measurement problem arises because the act of measurement fundamentally alters the state of a quantum system.

Unlike classical systems, wherein properties of an object such as the position and velocity exist without relation to observation, in a quantum system, an observation affects the state itself. This paradox is built upon the wave-particle duality of quantum objects whereby electrons, for example, have no definite properties about themselves, such as the position or momentum, except upon observation. Instead, these properties are described by a wave function, which encapsulates all the possible states the system could be in. Before measurement, a quantum system is said to exist in a superposition of all possible states, and the measurement causes the system to 'collapse' into a single state. This is starkly different from classical systems, where objects have definite states independent of observation.

Classical physics assumes that objects possess definite properties, such as the position of a car on a road or the temperature of a substance, which can be measured precisely without affecting the state of the object. Quantum systems, by contrast, are governed by the laws of quantum mechanics and exhibit a strange behaviour known as superposition, where a particle can be in multiple states simultaneously. These superpositions then collapse into one definite state upon measurement, but until then, the system is probabilistic in nature, described by the wave function. There is thus an inherent uncertainty in quantum measurement, in that the outcome is not determined until it is observed. This is encapsulated in Heisenberg's uncertainty principle, which suggests that certain properties, such as position and momentum, cannot be known simultaneously with absolute precision.

This further gives rise to debates on the role of the observer in quantum mechanics. Classical physics assumes that there is an objective reality which exists independently of observation. In contrast, some interpretations of quantum mechanics, like the Copenhagen interpretation, postulate that the process of measurement itself is the cause for the result of a quantum event. This challenges the classical notion of an objective reality and raises questions about the very nature of existence. Does reality exist independently of observation, or does it only come into being when observed? This philosophical dilemma, known as the measurement problem, continues to be one of the most intriguing and debated aspects of quantum mechanics. Even though quantum theory is well developed, the measurement problem still persists. There is no universal interpretation of quantum mechanics that would explain how the observation and measurement of a system would affect its state.

Uncertainty directly affects the performance of scientific

instruments. For example, STM relies on quantum tunnelling to image surfaces at the atomic level, where uncertainty in electron position and energy is crucial. Similarly, high resolution in electron microscopes is achieved by utilizing wave properties of electrons, but it is limited by the uncertainty principle. This is also reflected in spectroscopy, where the spectral line resolution is limited by the energy-time uncertainty relationship, thus affecting fields such as astrophysics and material science.

Uncertainty benefits modern technology as well. Transistors in a microchip are at nanoscales where quantum effects control the design and functionality. In quantum cryptography, uncertainty is used to ensure communication, since the act of measurement of a quantum state changes it, making eavesdropping detectable. GPS systems, which rely on atomic clocks, find that the accuracy of navigation is affected by the uncertainty in energy and time.

Uncertainty is also important in nature. Quantum tunnelling enables the particles to cross the potential energy barriers in nuclear fusion in stars, which makes possible the Sun's energy output. At the biological level, it determines the way electrons behave in atoms; it influences chemical bonding, photosynthesis, and how proteins fold.

Moreover, advanced sensors, such as those used in LIGO to detect gravitational waves, operate at quantum limits where uncertainty constrains their sensitivity. Even medical imaging techniques, such as MRI, are shaped by this principle, which affects the resolution of measurements.

Can Mankind Ever Know Reality?

Reality is layered and complex, therefore, existing beyond the reach of immediate perception. Scientific advancements have expanded our understanding, from the microscopic world of quantum particles to the vastness of the cosmos. Yet even the most sophisticated theories, whether quantum mechanics or general relativity, reveal limitations. The

BEYOND THE ATOMIC LEVEL

Heisenberg Uncertainty Principle and Gödel's Incompleteness Theorems suggest some inherent boundaries to what can be known or proven, emphasizing that some aspects of reality might forever lie beyond comprehension.

Human understanding is, by its very nature, a subjective thing. Perception shapes our understanding, and mental constructs such as language, mathematics, and culture filter the infinite complexity of existence into comprehensible frameworks. Powerful though these frameworks are, they cannot be said to capture reality itself.

Philosophically, reality could be infinitely vast or multidimensional, transcending human experience. Questions about the origin of existence, consciousness, and the ultimate purpose of the universe may remain unanswered due to the limits of human cognition. What if reality is not meant to be fully understood, but rather experienced?

Humanity's perception of reality can always be incomplete, not because of lack of effort but because of the infinite and subjective nature of existence. But the pursuit of knowledge remains profoundly meaningful in that it stimulates curiosity, creativity, and a deeper connection to the mysterious universe we inhabit.

Beyond science, philosophical questions about existence—such as why there is something rather than nothing, or what lies beyond the observable universe—may defy comprehension. Our cognitive faculties, evolved for survival on Earth, might not be equipped to grasp these transcendent truths. Even mathematics, the language of the cosmos, encounters paradoxes and incompleteness, hinting that some truths may forever escape logical deduction.

However, human beings have an incredible ability to abstract, imagine, and be creative. What was once thought to be incomprehensible, such as the curvature of space-time or the nature of quantum entanglement, has been made accessible through innovative thinking. This would imply that the unknowable is not a fixed point but a horizon that shifts with the advancement of knowledge.

However, the unknowable might not be fully reducible to human terms. It may represent a boundary where language, logic, and perception falter. Rather than fully comprehending these mysteries, we may learn to appreciate them, acknowledging that some aspects of reality are meant to inspire wonder rather than provide answers.

Scientific models represent the way the natural world behaves and behaves in the future. They are not reality itself. Models are abstractions, built with mathematics, logic, and empirical data to simplify and represent complex phenomena. Although they give a great amount of insight, they are enlightening and inherently limited in their relationship to reality.

Scientific models, by design, only capture specific aspects of reality. For example, Newtonian mechanics is a very successful model of motion at ordinary scales, but it fails in situations involving high speeds or strong gravitational fields, where Einstein's theory of relativity is required. Likewise, quantum mechanics models subatomic particles with fantastic accuracy, but it challenges classical notions of determinism and locality, demonstrating that reality at small scales behaves fundamentally differently than our intuition suggests. These examples highlight how models are context-dependent and evolve as our understanding deepens.

Models aim to be accurate while being bound by human perception and the tool we use to measure reality. They take the complexity of the universe into frameworks that can be comprehended by the human mind, but the translation process is fundamentally selective and therefore stresses certain elements while leaving others behind. For this reason, models cannot be taken as precise reflections of reality but are rather approximations, fitting in with the phenomena that are observed.

BEYOND THE ATOMIC LEVEL

Moreover, philosophical questions arise regarding the extent to which models reflect the "true" nature of reality. Do they represent reality as it is, or are they just useful constructs for organizing our observations? Some argue that scientific models, no matter how refined, are always provisional, subject to revision or replacement as new evidence emerges. This reflects the self-correcting nature of science but also underscores the limitations of models in fully capturing reality.

In this, we can find insight from old wisdom, like the Nasadiya Sūkta from the Rig Veda, which admits uncertainty and provisionality in understanding the universe: "There was neither existence nor non-existence then; there was neither the realm of space nor the sky beyond. Darkness was enveloped in darkness, and all was indistinguishable. Those who were engaged in deep meditation, who were the seers of the truth, found the unmanifested reality." This verse is a contemplation of the primordial state of the universe, wherein uncertainty reigned, and understanding was incomplete. It reminds that just as ancient musings of life, our scientific theories are but a provisional representation of a much deeper and inaccessible truth.

5: CAN SOMETHING BE NOTHING?

We may meet many paradoxical people in our lives. They seem good at first, but eventually, they reveal a different side, turning bad for us. How can they be both at the same time? This naturally makes us question their true nature. Now, imagine comparing them to particles. Just as these people display a dual personality, particles in the quantum world exhibit a similar duality. A particle manifests simultaneously the properties of both a wave and matter, bringing forth questions about what it is really.

This bizarre behaviour, termed wave-particle duality, throws open a new window to the quantum world and makes us question even the basis of reality. It's not about the particles but about the possibility of existence itself. These paradoxical people make us question the stability of our relationships. In the same manner, this dual nature of particles questions what we call "real."

And if particles can exist in dual states, meaning showing two sides at once, does this maybe hint that there is reality that we cannot perceive? This duality may, for instance, point towards parallel worlds, where multiple realities coexist, remaining elusive to us. The behaviour of particles teaches us not only about the quantum world but also inspires us to

BEYOND THE ATOMIC LEVEL

think beyond it, to imagine a universe far stranger than we ever believed possible. At the heart of our daily experience, nothing seems more fundamental than the notion that "something" is real, tangible, and existent, while "nothing" is the absence of all things.

Consider a quantum particle that can exist in two positions: position A and position B. According to superposition, the particle does not occupy just one of these positions before measurement; instead, it exists in a state where it is both at position A and position B simultaneously, with certain probabilities associated with each state.

In 1935, one of the pioneering fathers of quantum mechanics suggested a thought experiment which goes on to become one of the most iconic illustrations ever of the paradoxical nature of quantum mechanics. This was then known as Schrödinger's cat, a thought created to criticize and show all the strange implications of a quantum superposition and how any observation plays a role to determine the state of an object. Though purely theoretical, this scenario remains influential for debates both in physics and in philosophy that challenge the human conception of reality and observation.

Schrödinger's thought experiment commences with the assumption: a cat is locked within a sealed box containing, among other things, a radioactive atom, a Geiger counter, a vial of poison, and a hammer. The mechanism works in this manner: if the radioactive atom decays (a random quantum event), the Geiger counter detects it, triggers the hammer to break the vial of poison, thereby killing the cat. If the atom does not decay, the vial will not break, and the cat will not die. The atom decays by quantum mechanics, and so it's in a superposition of decayed and undecayed states until it is measured.

By extension, until the box is opened and the cat is observed, the cat itself exists in a superposition of two states: alive and dead. It is only when the box is opened and the cat is observed that the quantum superposition collapses into one definite state—alive or dead.

The thought experiment was Schrödinger's way to indicate the apparent absurdity of applying quantum mechanics on macroscopic objects. Since the principle of superposition is well established and experimentally tested for particles like electrons and photons, it is a challenge when this principle is extended to daily objects like a cat; something as tangible as a cat can really be both dead and alive at the same time?

The measurement problem in quantum mechanics is demonstrated by this paradox-the apparent fact that the act of observation plays a crucial role in determining the state of a quantum system. Quantum-mechanically, particles exist as a cloud of probabilities, described by a wave function, until they are measured. Schrödinger's cat forces us to confront an uncomfortable question: does this principle apply only to the microscopic world or does it govern the macroscopic reality that we experience daily?

Schrodinger's cat is much more than an attack on quantum mechanics but also a deep philosophical reflection on reality and observation. It places the chasm between the quantum and the classical in sharp relief. Whereas in the quantum world, events are probabilistic, governed by superposition, in the world of classical physics, events are deterministic and governed by the laws of Newton. The thought experiment blurs the lines between these two realms, suggesting that our classical understanding of reality may be nothing more than an approximation of a deeper, probabilistic quantum reality.

The experiment also raises the question of how much role the observer might play. In quantum mechanics, an observation is assumed to cause the wave function of a system to collapse into a definite state. But what constitutes an observer? Is it the human opening the box or could it be

BEYOND THE ATOMIC LEVEL

any sort of interaction with the environment that forces the system into some definite state? This question has been involved in some very interesting and philosophical debates, such as the Copenhagen interpretation and many-worlds interpretation of quantum mechanics.

Though Schrödinger intended the thought experiment as a critique of the Copenhagen interpretation, it has since been embraced as a powerful metaphor for the peculiarities of quantum mechanics. The concept has inspired various interpretations of quantum mechanics. For instance, in the many-worlds interpretation, the cat is neither alive nor dead; instead, the universe splits into two parallel worlds—one where the cat is alive and another where it is dead.

In practical terms, the thought experiment draws out the conflict of reconciling quantum mechanics with everyday experience. It further illustrates how our intuition for a classical world can only take us so far into the quantum realm. Experiments using quantum systems like qubits in quantum computers reference the principles of what was demonstrated in Schrödinger's cat about harnessing quantum superposition to create practical effects in real life.

The thought of Schrödinger's cat is more than a quirky experiment; it is a profound exploration of the boundaries between quantum mechanics and classical reality. It forces us to question the nature of existence, the role of observation, and the meaning of reality itself. Even though it was intended to critique quantum mechanics, the thought experiment has become a cornerstone for understanding the philosophical and scientific implications of the quantum world. In doing so, it has not only reshaped physics but also inspired a broader reflection on the mysteries of the universe.

Superposition is the principal principle in quantum computing that allows quantum bits, or qubits, to exist in multiple states at the same time as opposed to classical bits restricted to either 0 or 1. This means a quantum computer can do many more calculations in parallel. As a result, quantum computers can solve certain problems exponentially faster than classical computers, which would take an impractically long time to compute using traditional methods. For instance, in tasks like factoring large numbers or searching unsorted databases, superposition allows quantum computers to explore all possible solutions at once, dramatically speeding up the process. This parallelism, among other quantum phenomena such as entanglement, allows quantum computers to solve complex problems in such areas as cryptography, drug discovery, and optimization in which classical computing struggles. The power of superposition makes quantum computing so potentially transformative - a new way to process information that classical computers can't replicate.

We might have to go to the cinema theater alone, so we call our friends to come with us. But they seem very busy and kindly decline. Later, we see them at the same theater with their own friends. Similarly, particles also appear and disappear in places, and we don't always know why.

Imagine you were standing at the base of a tall mountain, and in classical physics, you would have had to climb over the mountain to get to the other side. If you had not had enough energy to reach the top of the mountain, you would be stuck here at the base.

However, in the quantum world, it is as if you have a magical ability to pass through the mountain without climbing it—appearing on the other side without ever traversing the space in between. In this analogy, the mountain is the potential barrier and the ability to "tunnel" through it, rather than climbing over it, reflects how quantum particles can pass through barriers that they seem to not have enough energy to overcome. Just like the person appearing on the other side of the mountain without climbing it, a quantum particle can appear on the other side of a potential barrier, even though it doesn't have the classical energy to get over it.

Quantum tunnelling is a very interesting phenomenon that defies our classical understanding of physics. It is the ability of particles to pass through potential barriers, which according to classical mechanics, should be impenetrable. In other words, in classical terms, a particle must have sufficient energy to overcome a barrier. If it doesn't, it would simply be reflected back. However, particles behave differently in the quantum world. Due to the wave-like nature of the particles, there is a non-zero probability that the particle may appear on the other side of the barrier without having crossed it in the conventional sense. This effect is one of the fundamental aspects of quantum mechanics and holds great significance for understanding nature and the development of modern technology.

Quantum tunnelling has been seen in many scenarios and plays a critical role in some of the universe's and human innovation's most crucial processes. Some of the most famous examples of quantum tunnelling include nuclear fusion-the process that powers the sun-and other similar types of stars. In the heart of a star, temperatures and pressures are enormous, but the energy of the nuclei in collision is frequently not enough to overcome the strong electrostatic repulsion between them. According to classical physics, fusion cannot occur under such conditions. However, quantum tunnelling allows them to avoid the energy wall, enabling fusion to continue and giving rise to a release of energy that forms the star and provides that light and heat which supports the life on earth.

Besides its cosmological importance, quantum tunnelling is also vital in modern electronics. Devices such as the tunnel diode depend on the phenomenon of tunnelling for their functionality. The tunnel diodes, which are used to allow electrons to tunnel through barriers in certain specific ways, are applied in high-frequency oscillators and amplifiers. With the miniaturization of electronic devices, the tunnelling effects are even more apparent and have a very crucial role in the operation of transistors in nanoscale circuits. This carries vast implications in the designing of faster and more efficient computing technologies.

The STM scanning tunnelling microscope was also one of the very notable applications of quantum tunneling. It is actually a device that makes use of the tunnelling effect for scanning surfaces at atomic scales. Bringing a sharp metal tip close to a surface to measure the tunnelling current flowing between the tip and the surface, scientists mapped the arrangement of individual atoms. This has completely altered the study and manipulation possibilities at the smallest scales of matter and has opened up unprecedented new opportunities in nanotechnology and materials science.

Tunnelling benefits quantum computation. In those advanced devices, tunnelling enables a particle such as an electron to pass through between its quantum states in ways the classical system could not account for. That is an aspect that makes the quantum computers compute at the level which quantum computers manage to tackle due to such principles, among others: superposition, entanglement, and tunneling.

Moreover, quantum tunnelling inspires deeper philosophical inquiry into what constitutes reality. This idea is counterintuitive: that a particle can come out of the other side of a barrier without crossing the intervening space. This makes our classical notions of causality and determinism somewhat hard to swallow, forces us to rethink the limits of what is possible, and challenges scientists to consider the probabilistic nature of the quantum world. The implications of tunnelling extend beyond physics into broader discussions about how we perceive and interpret reality.

It is not just a theoretical curiosity; it is a phenomenon

with real-world significance that influences both our understanding of the universe and the technologies that shape our lives. From powering the stars to enabling advancements in electronics, microscopy, and computing, tunnelling exemplifies the strange and wonderful nature of the quantum realm. It bridges the gap between what is seemingly impossible and physically realizable, hence manifesting the power of quantum mechanics to challenge and broaden the limits of human knowledge.

The concept of "nothingness" in quantum mechanics is very much tied with the probabilistic nature of particles and their behaviour. Unlike classical physics, where particles are definite in position and occupy definite locations, quantum mechanics introduces the concept of probability amplitudes, which describe the likelihood of finding a particle in a given region. In this framework, "nothingness" can be thought of as regions where the probability amplitude is zero or nearly zero, indicating that the particle has an extremely low chance-or no chance at all-of existing there.

This probabilistic view challenges the classical idea of an absolute void. In quantum mechanics, even "empty" space is not really empty but filled with fluctuating fields and virtual particles that momentarily come into existence due to quantum fluctuations. However, when examining specific regions in terms of probability amplitudes, the possibility arises that a particle might not exist at all in those areas. It means, therefore, that there could exist such partial or conditional "nothingness," where the lack of a particle is not fixed but rather probabilistic.

This has a significant consequence. To take one simple example, a quantum particle could tunnel through a barrier from the left even if there were zero probability for the particle to exist anywhere in the barrier region on the left. Again, depending upon the interpretation, it might happen that whether the particle was in the region or not was decided only at measurement. The state of the particle can be expressed until measured as a superposition of probabilities, even though in some regions it doesn't exist at all.

By framing "nothingness" in terms of probability amplitudes, quantum mechanics draws the line between existence and non-existence. This concept does not only change our vision of physical reality but also brings up philosophical discussions regarding being and the role of observation in defining reality itself. In the quantum world, "nothingness" is not an absolute absence but a dynamic and probabilistic state, reflecting the uncertain and non-deterministic nature of the universe.

At the heart of modern physics lies quantum field theory (QFT), which, building on quantum mechanics and special relativity, presents the world of fundamental particle interactions. According to QFT, the universe does not consist of particles per se, but is instead characterized by fields that fill the entirety of space. The most significant aspect of the so-called "vacuum state" is that even in this energy level state, it is far from being empty. Instead, they are dynamic, seething with energy and activity. Quantum fluctuations in these fields give rise to virtual particles, fleeting entities that momentarily appear and disappear, challenging our conventional notions of emptiness and existence.

Virtual particles are not directly observable but have measurable effects, such as in the Casimir effect or Hawking radiation near black holes. The Casimir effect occurs when two uncharged, closely spaced metal plates in a vacuum experience an attractive force due to fluctuations in the quantum field. This phenomenon demonstrates that "empty" space is far from empty; it is alive with activity at the quantum level. Similarly, Hawking radiation arises because quantum fluctuations near the event horizon of a black hole allow virtual particles to become real, which gradually causes the black hole to evaporate itself. These examples thus emphasize that the idea of a true vacuum-an utterly empty space devoid of any activity-is incompatible with our current understanding of quantum field theory.

The presence of quantum fluctuations gives rise to deep philosophical questions on the nature of nothingness. What we consider as "nothing"—an absence of matter and energy—is, in fact, a dynamic field of potential and possibility. Quantum fluctuations permit energy to momentarily "borrow" from the vacuum, allowing the creation of pairs of virtual particles that annihilate each other almost instantaneously. These fluctuations are intrinsic to the quantum fields that underlie all of reality, suggesting that "nothingness" as an absolute state does not exist in nature.

This understanding leads to a critical question: Is true nothingness possible, or is it merely a human construct, tied to our limited perceptions and linguistic frameworks? In classical physics, nothingness might be defined as an absence of particles, forces, and interactions. However, quantum mechanics has revealed fields and fluctuations continue to remain even in the absence of observable matter. Nothingness, as such, then is quite relative, scale-dependent as well as frame-dependent too.

The rejection of true nothingness has profound implications for philosophy. It challenges deeply ingrained human concepts of existence and non-existence, compelling us to rethink the lines between the two. If even the vacuum is chock-full of energy and potential, then what is it to "exist"? Does the fleeting appearance of virtual particles in the vacuum count as existence, or are they merely mathematical artifacts of our theories?

It is also consistent with some philosophical traditions that nothingness cannot be present in nature. For instance, in eastern philosophies such as Buddhism, the concept of śūnyatā - emptiness - does not imply a void but refers to an interdependent existence in which everything arises because of conditions and relationships. Similarly, quantum field theory tells us that even vacuum is interrelated with everything else in the universe: full of potential for creations and transformations.

The interplay between quantum field theory and philosophical ideas on nothingness underlines the limitations of human perception and language. Our senses and tools are designed for a macroscopic world, where emptiness is absolute and tangible. Yet, when we push ourselves into the quantum regime, we find that intuitive concepts break down. The very fabric of the universe, once composed of discrete building blocks, is now a dynamic and probabilistic tapestry of fields and fluctuations.

This raises profound questions about the nature of reality. If nothingness is unattainable, then what lies at the foundation of existence? Is the universe inherently full, its apparent emptiness an illusion born of our macroscopic viewpoint? Or is our inability to observe true nothingness simply a limitation of our current understanding, leaving room for future theories to explore deeper truths?

Quantum field theory fundamentally redefines our understanding of nothingness. It was once thought of as space devoid of all things. Now we know it as a sea of activity governed by quantum mechanics. Virtual particles, quantum fluctuations, and omnipresent energy in the vacuum challenge the notion of voids in the classical world and ask us to reevaluate existence. This helps further unravel the mystery of the quantum world while still slowly nudging us closer to solutions of age-old questions philosophers ask about the bounds of human understanding and reality as a whole. This constant play between science and philosophy shows us that the universe is far more intricate and mysterious than any refined theories might yet allow.

The strange and counterintuitive quantum world

profoundly challenges our classical perceptions of reality as well as the concept of nothingness. In classical thinking, reality is fairly simple: objects exist at well-defined locations, space can be empty—a perfect void in which no matter or energy exists—and everything is determinate. Thus, a pure absence-nothing—is an absolute zero, without any interactions or particles that may have given rise to forces. However, the quantum world breaks all these assumptions; here is a reality far more dynamic, interconnected, and puzzling than even a classical physicist could imagine.

Quantum mechanics also reveals phenomena such as superposition, entanglement, and tunnelling that classify particles behaving in ways inconceivable to a classical mind. A particle could exist in several states at one time, interact with another particle instantly over vast distances, or appear on the far side of an energy barrier without crossing it. All of these behaviours violated classical conceptions of separate and independent objects and showed a reality that was basically a web of probabilities deep inside observation and measurement.

Even more striking is the quantum field theory perspective, where the vacuum itself—the very embodiment of nothingness in classical physics—is revealed to be seething with activity. Quantum fluctuations constantly bring virtual particles into and out of existence, ensuring that "empty" space is never truly empty. This innate dynamics of the vacuum goes against the postulate of absolute void; rather, nothingness should be considered relative and conditional in nature, conditioned upon the quantum fields that fill the universe.

This quantum redefinition of reality poses some profound implications for our perception of existence. If it is true that particles and fields are governed by probability and uncertainty, and even that the vacuum teems with energy and activity, what does it mean to "exist"? Must that exist be permanent, or is it possible for something to exist in a way that is fleeting and probabilistic? Is nothingness a physical reality, or is it more a concept that arises out of human perception and language?

Philosophical questions abound arising from these quantum revelations that touch the human mind beyond itself. They challenge the status quo and make one review the concept of being and the role of observation on reality as well as those limits set by humanity upon its understanding. Quantum mechanics, in its complexity through mathematics and the mysteries in phenomena, reminds man of the operation of the universe on principles that often belie his daily experiences. It forces us to face the prospect that our classical understanding of the world—built on concepts of certainty, determinism, and a dualism of being and nonbeing—is at best an approximation of a far stranger, more complex reality.

It, in this sense, makes the quantum world not just a domain of scientific inquiry but also a source of profound philosophical insight. It challenges us to embrace uncertainty, to see reality not as a static and deterministic construct but as a dynamic and probabilistic interplay of possibilities. It blurs the line between existence and nothingness, suggesting that what we perceive as empty or absent may, in fact be teeming with hidden potential.

Ultimately, the strange quantum world forces us to expand our understanding of reality and confront the limitations of our classical perceptions. It invites us to explore not just the physical nature of the universe but also deeper philosophical questions about what it means to exist and whether true nothingness is ever possible. In doing so, it highlights the profound mystery of the cosmos, reminding us that in the quest to understand reality, the journey of the mind is as much about the pursuit of scientific knowledge. The quantum world is not merely a challenge to our assumptions but transforms them, opening new windows to the nature of existence and the infinite complexity of the universe.

6: ARE WE ALL CONNECTED?

Lena sat on her balcony in Paris, the evening air cool against her skin. She sat there holding a notebook on her lap, pen poised over the paper, ready to capture something that seemed elusive. Her mind was a maelstrom of thoughts about Noah, who was thousands of miles away in Tokyo. They had chanced to meet once at a long-forgotten meeting in Rome years earlier, and though life continued to tear them apart, the connection was there, in the unsaid: this invisible string that connected their lives.

Tonight, of all nights, she needed it the most.

In the Tokyo apartment, Noah is sitting at his desk as his fingers graze upon a journal he hadn't opened in months. He didn't know what had brought him to the page tonight, but when it opened, memories of Lena began to flood into his mind. He started to write without knowing why; just let the words go, as if something bigger than himself was guiding them. And at that moment, on the other side of the world, Lena sat across from him, writing, her pen moving with thoughts of Noah.

Neither knew the other was writing, yet their words mirrored one another—fragmented memories of Rome, shared laughter, fleeting gazes. It was as if some unnameable force was orchestrating their actions, making them one despite the distance.

According to classical physics, this connection could be explained as a coincidence, as two individuals happened to think of each other at the same time. But in quantum mechanics, their interaction could be viewed differently. Consider them as entangled particles. In quantum entanglement, the state of one particle instantly affects the state of another, no matter how far apart they are. Just as their words mirrored each other, perhaps their connection wasn't just a coincidence—it could be a manifestation of entanglement, an invisible link that transcends the limits of space and time.

It has always been one of the prime concerns for scientists and philosophers, as human understanding and perception of the cosmos changes with time. In the old days of classical physics, the universe was mainly seen as consisting of disparate and separate things. This view of objects was an independent and separate system because of the existence of different forces such as gravity or electromagnetism, which exerted influence over them. This framework stressed individuality and separateness, depicting the universe as an enormous mechanical system in which interactions occurred by direct contact or through measurable influences over distance. For example, planets orbit the sun due to gravitational forces, but within this view, they are still distinct bodies with predictable paths.

This classical view was soon challenged by the advent of quantum mechanics at the start of the 20th century. Quantum mechanics revealed a fundamentally different picture of reality, one that undermined the notion of strict separateness between entities. At the quantum level, particles exhibit wave-particle duality, existing not as solid, discrete objects but as probabilities until observed or measured. This principle of superposition, where particles can exist in multiple states simultaneously, illustrates a level of interconnectedness that classical physics could not accommodate. In addition, the non-locality of quantum mechanics tells us that the motion of one particle can affect another instantly, even at such huge distances. The quantum effect then breaks with classical causality and claims that particles are a single system, rather than independent units.

In quantum entanglement, it is found that once two particles are entangled, their quantum states become completely dependent on each other. The measurement of the state of one particle at once determines the state of the other, with the physical distance between the particles being irrelevant. The strength of this connection profoundly challenged the traditional understanding of space and time because it indicated information exchange instantaneously across seemingly impossibly large distances. In support of these correlations and not explainable by either classical theories or hidden variables is evidence from experiments, such as results from Bell test experiments. Instead, this provides evidence for a deeper connection, or intrinsic relationship among particles. This implies something like unity within the world itself.

However, quantum mechanics paints a very different picture of the universe. According to quantum theory, particles can be entangled in such a way that their states become correlated instantaneously, regardless of the distance separating them. This phenomenon is known as quantum nonlocality, which denies the classical notions of locality, suggesting that particles may influence each other across vast distances without any time delay. It is a phenomenon where particles that are entangled can influence one another instantaneously, regardless of the distance between them. This instantaneous connection seems to violate the very basic principle of locality in classical physics, which dictates that objects can only be influenced by their immediate surroundings, and the effects propagate at a finite speed, never faster than the speed of light.

In classical physics, space and time are separate entities which cannot change. In classical physics, objects exist in definite points in space and time; interactions of the objects happen at specific places and according to definite sequence. However, with the advent of quantum mechanics, a different paradigm begins. Particles, specially those that are entangled, do not obey such laws. Entangled particles have created a special bond where the attributes of the particles are somehow correlated in ways that go beyond the capability of localized interactions. If two particles are entangled, for instance, the state change of one particle instantaneously influences the state of another, regardless of how large the distance between them. This phenomenon, observed in experiments such as the famous Bell's Theorem, shows that the behaviour of entangled particles is not confined to the classical idea of locality, where influences must travel through space at a finite speed.

The notion of nonlocality challenges our understanding of how the universe operates. In classical mechanics, the world operates under the assumption that cause and effect are tied to proximity. If something happens to an object, then the effect must occur within some spatial and temporal context. Quantum nonlocality seems to suggest that particles are able to act as if they are "aware" of each other at huge distances, in ways which bypass the restrictions placed upon them by classical theories of space and time. This leads to a paradox since it seems to imply information could travel faster than light, or even more weirdly, that space and time may not be as fundamental as we have assumed.

Quantum entanglement plays a central role in this phenomenon. Two particles that become entangled cease to have independent quantum states. The state of one particle directly determines the state of the other, even if they are light-years apart. This link, famously referred to by Albert Einstein as "spooky action at a distance," implies that the particles are connected in a way that defies any local explanation. Thus, the entangled particles aren't two different objects but belong to a single system with information about one particle innately contained in the other, irrespective of the distances between them. This attribute of quantum mechanics makes it difficult to understand non-locality as an important ingredient of entanglement itself and is very hard to connect it to any intuitively and classically formed view of the world.

The most puzzling aspect of quantum nonlocality is that it does not allow any faster-than-light communication or violation of causality, despite appearances to the contrary. The influence between entangled particles happens instantaneously, but no usable information can be transmitted between them faster than light. This implies that even though the states of the particles are instantaneously correlated, this has no paradoxical implications at all regarding communication or causality. However, the difference does not reduce the weird and counterintuitive character of the connection which quantum nonlocality seems to represent.

Bell's theorem, or simply Bell's inequality for a shorter term, is a principle, first proposed by John Bell in 1964, in which quantum mechanical reality is probed and is based on the limits placed on classical physics. A no local hidden variable theory-that is, a local theory with realism, because it says that particles have definite properties quite independently of measurement-can consistently reproduce the statistical predictions of quantum mechanics. In essence, Bell's theorem demonstrates the fact that if quantum mechanics is valid, the world cannot be as imagined within the frameworks of classical physics. Experimentally speaking, the significance of the theorem is in it providing the opportunity to find out if the correlations suggested by quantum mechanics are fundamental or a local hidden reality dictates what the particles are doing. Since then, many experiments based on Bell's theorem have been performed, and the results overwhelmingly support the quantum mechanical view, indicating that particles can indeed behave nonlocally, where their states are correlated over vast distances instantaneously. This has profound implications for understanding the nature of the universe, suggesting that reality at the quantum level is inherently interconnected in ways that defy classical notions of space, time, and causality.

In classical physics, the concept of hidden variables plays a very important role in the underlying assumptions about how physical systems operate and how the properties of objects are determined. Hidden variables refer to hypothetical factors that are not directly observed or measured but are believed to influence the behaviour of a system. In classical mechanics, one assumes the existence of hidden variables as a larger belief that the world follows the deterministic and predictable principles by well-defined laws of nature.

In classical mechanics, all physical objects are presumed to have definite properties irrespective of observation. For instance, for a moving ball, at any time, it would be possible to have its position, velocity, and momentum known precisely. If we knew everything about all the particles and forces in a system, classical physics would predict that we could know the future behaviour of that system with absolute certainty. The behaviour of each individual particle, as well as the interactions between them, can be described through deterministic equations of motion, such as Newton's laws.

The concept of hidden variables arises from the idea that, in principle, if we knew the complete state of a system-down to the finest details of each particle's position, velocity, and other fundamental properties-then we could predict all future events with perfect accuracy. The assumption is that these hidden variables are the underlying factors that determine the state of a system, even though they are not directly observable or measurable with current technology. Classical physics suggests that these hidden variables do not violate any fundamental laws of physics; instead, they represent additional information that would allow for a complete and precise description of the system.

By having additional unseen or hidden variables, quantum versions help explain phenomena that seems more random or probabilistic with a classical statistical mechanics in nature. For instance, whenever dealing with large populations of particles, such as a gas or fluid, seemingly unpredictable or chaotic motion arises by the individual particles; the classical physics assumes just our inability to observe fully factors affecting the behaviour for a particle. Hidden variables would, in that sense, explain the unpredictability, with the actual state of each particle being determined by exact, underlying variables, which are simply not available to us in practice.

One of the most important aspects of hidden variables in classical physics is the principle of determinism. Determinism suggests that the future of any system is entirely dependent on its present state, and if we had access to all the hidden variables, we could predict the future behaviour of that system with no uncertainty. This deterministic view of the universe has been central to classical thought, from the work of Isaac Newton to the development of electromagnetism by James Clerk Maxwell and the kinetic theory of gases. In this worldview, the apparent unpredictability of certain phenomena is not a fundamental feature of reality but rather a result of our inability to measure all relevant variables.

However, this classical thought was overthrown with the emergence of quantum mechanics in the early 20th century.

Quantum mechanics, that describes the behaviour of particles at a microscopic level, was a highly different approach in understanding reality. Unlike the classical approach, quantum mechanics includes intrinsic uncertainties in measuring certain properties, such as the position and momentum of particles. These uncertainties are not due to lack of information but are, rather, fundamental features of the quantum world. The idea that particles do not have definite properties until measured—the sort of thing that quantum mechanics revealed through concepts like wave-particle duality and superposition—contradicted the deterministic worldview that classical physics had held for centuries.

This led to the development of hidden variable theories in quantum mechanics, as some physicists - notably Albert Einstein - felt that this probabilistic nature may just be due to other, hidden variables we might not have observed yet. As they saw it, apparent randomness in quantum events should have an underlying determinism; the hidden variables must have determined the outcome of the measurements, but must have been inaccessible to us.

Bell's theorem proved, however, that no local hidden variable theory could reproduce all the predictions of quantum mechanics. This meant that a kind of randomness observed for particles at the quantum level couldn't be attributed to classical hidden variables. Quantum mechanics brought a new paradigm with a fundamentally indeterminate character of particle behaviour governed by probabilities rather than by determinist laws. The experimental verification of Bell's theorem further confirmed that quantum mechanics is correct and that any hidden variables must be nonlocal, meaning they cannot adhere to the constraints of classical locality.

The number of seminal experiments that test the predictions of Bell's theorem along with the observation that quantum behaviour should not be attributed to some local hidden variables and has a true, genuine manifestation of nonlocal behaviour have proven of crucial importance for validating evidence about the soundness and validity of quantum mechanics departing from classical locality realism and classical concepts of reality. Some of the most impressive experiments include those reported by Alain Aspect, John Bell, among others, thus adding more evidence to establish quantum nonlocality.

One of the best experiments testing Bell's theorem was performed by Alain Aspect and his team in 1982. Aspect's test proved to be a direct measure of the violation of inequalities by Bell, which are certain inequalities that any local hidden variable theory must satisfy. Measuring correlations between pairs of entangled photons through polarization was the experiment.

Aspect's apparatus consisted of a source that emitted pairs of photons in an entangled state. The photons were sent to two detectors, which were spatially separated by a large distance. The crucial feature of the experiment was that the settings of the measurements (the angles at which the polarizations of the photons were measured) were chosen randomly when the detection occurred, so the measurements were spacelike separated. This ensured that no signal could propagate between the detectors faster than the speed of light, thus keeping the local hidden variable theories from affecting the outcome.

The results of Aspect's experiment demonstrated that the correlation between the polarizations of photons did indeed violate the Bell inequalities, which was exactly as predicted by quantum mechanics if the photons were entangled with one another and influenced one another instantaneously, no matter how far apart they are. This experimental violation of Bell's inequalities was quite an evidence for quantum non-locality and ruled out all other local hidden variable predictions.

A very interesting experiment that is known as the "loophole-free" Bell test was conducted in 2015 by a team of researchers led by Ronald Hanson from Delft University of Technology. This experiment was particularly important because it addressed two major loopholes in previous Bell test experiments: the "locality loophole" and the "detection loophole."

The locality loophole arises from the fact that previous experiments could not guarantee that the measurement settings on both sides of the experiment were chosen independently and randomly. If the measurement settings were predetermined or influenced by local hidden variables, the results could have been biased. The detection loophole arises from the fact that in some experiments, not all entangled particles were detected, which could have affected the statistical results.

The measurement settings of 2015 experiment were selected using a new technique to ensure their independence, and a detection method that included nearly all the entangled particles. Consistent with quantum mechanics predictions, the results of this experiment clearly violated Bell's inequalities. This was the most convincing evidence yet that it was quantum mechanics, not some local hidden variable theory, which governed the behaviour of the entangled particles.

Another important experiment was conducted by Nicolas Tittel and his collaborators in 1989. This experiment was a test of Bell's theorem, but this time with a different approach. The purpose of the experiment was to test the violation of Bell's inequalities in a situation where the two measurement stations were separated by a considerable distance. A significant feature of the experiment is its application of a "delayed-choice" scheme where it was decided that when a photon would be emitted that particular setup of measurement devices has been made. They looked for correlations which could not be explained as arising from communication between the measurement stations, and thus sought to eliminate the possibility of the photon "knowing" which way to measure before it was detected. The results showed a clear violation of Bell's inequalities and further confirmed the idea that entangled particles can exhibit instantaneous, nonlocal correlations.

In 1998, a team led by Thomas Weihs conducted another important experiment testing the violation of Bell's inequalities.

This experiment, performed at the University of Innsbruck, involved entangled photons produced by a calcium atom and sent through optical fibers to two detectors placed a significant distance apart. The measurement settings were chosen randomly on the detectors so that no signal traveled between the detectors faster than the speed of light, thus any communication that could allow a local hidden variable to be involved in determining the result was prevented. Again, this experiment confirmed all predictions by quantum mechanics that the violation of Bell's inequalities among correlations in photons' polarizations would appear. Some gaps from the earlier experiments were thereby filled in through the successful realization of this experiment: It supported evidence for nonlocality and quantum entanglement being reality.

A third experiment was done by Maria Giustina of the University of Vienna with her research group in 2010.

It made use of an entangled photon source, where both the locality and the detection loophole would be eliminated from a potential problem of the earlier experiment. The researchers used a random number generator to ensure that their choice of measurement settings did not depend on any kind of hidden variables, and they selected high-efficiency detectors for the purpose of ensuring maximum detection of entangled particles. Results for the Giustina experiment have yet again provided violation of Bell's inequality with measured correlations beyond possible explanation of the theory that supports the concept of local hidden variable theories. Such a results further added strength to evidence building to the prediction from quantum mechanics and their outcomes in exhibiting non-local behaviour.

The series of experiments testing Bell's theorem, from Aspect's seminal work in 1982 to the loophole-free tests in 2015, have provided overwhelming evidence that quantum mechanics accurately describes the behaviour of entangled particles and that nonlocality is a real and inherent feature of quantum systems.

These experiments have profoundly influenced our view of the universe, challenging the classical views of space, time, and causality and demonstrating that, at the quantum level, particles are interconnected in ways that elude the strictures of classical physics. The persistent violation of Bell's inequalities in these experiments continues to be one of the most compelling demonstrations of the strange, nonlocal nature of the quantum world.

In Bell's theorem, there is great implication in the reality aspect, especially in terms of how we perceive the connection between the microscopic quantum particle world and the world as we experience it in a macroscopic scale. On a very basic level, Bell's theorem argues that the behaviour of quantum systems cannot be explained using local hidden variables. The theorem shows that quantum mechanics predicts phenomena that are not to be found classically, offering a glimpse of reality far more interconnected and nonlocal than anyone previously imagined.

Bell's theorem was a mathematical breakthrough which made the tension between quantum mechanical predictions and local realistic principles clear, an important idea in classical physics. Local realism is the view that objects have definite properties independent of observation, but information about these properties can only be transmitted at or slower than the speed of light. Bell derived a set of inequalities, now called Bell's inequalities, that local hidden variable theories satisfy; however, quantum mechanics predicts correlations between entangled particles that violate these inequalities and thus cannot be reconciled with local realism.

Experimental verification of the theorem came in the form of few groundbreaking experiments, notably from Alain Aspect's in 1982 and also in loophole-free tests later in 2015, which measured correlations between particles, such as photons in pairs, that were created in entangled states sent in opposite directions. The key feature of these experiments is the random selection of measurement settings, which guarantees space-like separation between the measurements, such that no signal can travel from one detector to the other faster than light. If local hidden variables were involved, then there would be settings for local hidden variables that would drive the results, but instead, the experiments clearly measured violations of Bell's inequalities, which showed that quantum mechanics correctly predicted the types of correlations observed and weren't accounted for by local theory.

The experimental results do confirm that quantum particles possess a kind of interconnectedness, which is sometimes termed "quantum nonlocality." It indicates that the state of one particle can be instantly affected by the state of another, even when separated by great distances. The very notion of nonlocality, which seems to imply that information travels faster than light, defies classical intuitions about space and time and challenges the concept of locality that is fundamental to classical physics. These experiments show that at the quantum level, particles do not behave as independent entities with predefined properties but are entangled in a way that links their states in a nonlocal, instantaneous manner.

Bell's theorem points to a deeper, interconnected reality that fundamentally alters our understanding of the nature of the universe. Before quantum mechanics, the world was largely understood through a classical lens, where objects had well-defined properties that could be measured and understood in isolation. In classical physics, reality was deterministic, and the behaviour of individual objects could be predicted with accuracy, assuming all forces and factors were known. However, quantum mechanics suggests that this view of reality is incomplete, at least through the lens of Bell's theorem.

Quantum level: The properties of particles are not well-defined until they are measured. This is not because they are unknown but rather an intrinsic feature of quantum systems. When two particles are entangled, their properties become correlated in ways that cannot be explained by classical theories. This immediately determines the value of the other particle's corresponding property, no matter how far apart the two are. The fact that the communication between the particles takes place instantaneously, that is faster than the light speed, and also contradicts the concept that space and time were taken as absolute, which were characteristics in classical physics.

Bell's theorem and interconnectedness, which is being suggested by this, leads us toward a reality fundamentally different from what classical physics would have us believe. The state of a system is, in quantum mechanics, more than a collection of properties, but rather something that is holistic and can never be described by summing its parts. The very nature of this interrelation hints at the idea that the universe is not merely an aggregate of separate, independent objects but a rather delicate network of relationships. Therefore, properties of individual particles cannot be separated from the whole. Bell's theorem reveals a more fundamental, nonlocal relationship between particles that go beyond the space and time horizon; it suggests that reality is much more complex and united than classical physics could account for.

Furthermore, such interconnectedness gives rise to profound philosophical questions on the nature of causality, free will, and what is possible for humanity to know. It also has a notion of influencing other particles instantaneously and independent of distance; this shows that the universe acts differently from our classic conception of cause and effect. It allows events in the quantum realm to correlate with one another in seemingly impossible ways in terms of space and time, breaking the conventional sense of intuition over how things go in this world.

From these suggestions, quantum nonlocality, via Bell's theorem and experiments verifying it, implies a probabilistic world, much more than being deterministic in nature. A world wherein the future of a given system can always be accurately predicted from a knowledge of its present was the worldview of classical physics. The quantum world does it with probability. This probabilistic nature of quantum systems suggests there is a fundamental limit to how much we can know about the universe and some elements of reality are ultimately unknowable. Measurement itself-the act that causes a quantum system to collapse to one definite state further pushes into view the role of the observer in the definition of reality, further adding another level of complexity to our perception of reality.

The term "spooky action at a distance," coined by Albert Einstein, summarizes the curious nature of quantum entanglement and the related phenomenon known as nonlocality. Einstein coined this expression after the strange, paradoxically behaving entangled particles were described in quantum mechanics. Spooky action" is the term used when it was discovered that quantum mechanics allows two or more particles to become entangled in a way that, regardless of the distance between them, the state of one particle is instantaneously correlated with the state of another. This concept shakes the very foundation of how we understand physical interactions in the universe because it fundamentally challenges classical notions of space, time, and causality.

The laws of nature, in classical physics, are governed by principles local and deterministic. Local theory, one of the cornerstone postulates of classical physics, stipulates that an object cannot be influenced by things located far away, for nothing travels faster than it to mediate that influence. This implies that information, energy, or influence cannot travel faster than the speed of light according to Einstein's theory of relativity. In this framework, every physical interaction must have a well-defined cause that occurs in a specific location at a specific time.

The concept of "spooky action" contradicts this view directly. Quantum entanglement, as illustrated by the phenomenon, is when two particles are entangled such that their properties—polarization or spin, for example—are linked in a manner independent of space and time. The state of one particle determines immediately the state of the other, regardless of distance. In other words, a measurement done on a particle in one place can immediately affect the state of a particle in another, distant place. Einstein famously attacked this phenomenon, calling it "spooky," because it seemed to indicate that the universe was acting in a way that was incompatible with classical notions of causality and locality.

This "spooky action" means that quantum mechanics cannot be comprehended in terms of classical physics. In classical thinking, the universe is a collection of independent objects whose states can be determined through direct interactions. The idea that two entangled particles can influence each other instantly—without any signal traveling through space—challenges the assumption that objects have independent and separable properties. This raised significant philosophical and scientific questions about the nature of reality, the role of measurement, and the limits of human knowledge.

Einstein's aversion towards quantum entanglement and "spooky action" was further fueled by the development of the EPR (Einstein-Podolsky-Rosen) paradox in 1935. The paradox was designed to highlight what Einstein considered the incompleteness of quantum mechanics. According to quantum mechanics, particles can be entangled in such a way that their properties are undefined until measured. However, if two entangled particles are separated by a large distance and one is measured, the outcome should instantaneously affect the other particle, regardless of the distance between them. Einstein, Podolsky, and Rosen argued that this "instantaneous" interaction seemed to violate the principle of locality and suggested that quantum mechanics was incomplete, and some hidden variables or deeper theory were missing.

In the 1960s, physicist John Bell formulated Bell's theorem, which mathematically demonstrated that no local hidden variable theory—one that maintains the principle of locality—could explain the experimental outcomes predicted by quantum mechanics. Bell's inequalities provided a testable distinction between quantum predictions and those of local hidden variables. If the predictions of quantum mechanics were correct, then no local hidden variable theory could explain the observed correlations between entangled particles. This was followed by a series of experiments, most notably Alain Aspect's experiment in 1982, that confirmed that quantum mechanics was correct and that the results violated Bell's inequalities.

These experiments demonstrated that "spooky action at a distance" was not just theoretical, but a real property that could be experimentally validated, namely that the quantum systems had a feature about it. The state of one particle was instantaneously decided by the measurement even though the particles were separated across enormous distances. This did therefore show that the localization aspect of the classical domain doesn't apply in this domain.

Quantum nonlocality requires a rethinking of some of the most deeply held assumptions in classical physics as its experimental verification is done. The implications of "spooky action" are profound, not only for quantum mechanics but also for the way we perceive the structure of reality itself. Among the most important consequences of quantum entanglement is the abandonment of the principle of locality, one of the core ideas in classical physics. We suppose, in the classical world, that objects are distinct, and their interactions are space and time limited. However, quantum mechanics shows us that objects can be profoundly connected, regardless of how much distance separates them, thereby challenging our classical view of space as a passive background upon which objects exist independently of each other.

The concept of spooky action has also been associated with another question: causality in quantum mechanics. In the classical world, we expect an evident cause-and-effect relation that operates within the frame of space-time. But in quantum mechanics, instant correlations between entangled particles do not fit into a chain of causality in a classical sense. This might mean that events in the quantum world may not proceed according to the deterministic structure of cause and effect that characterizes classical physics.

Quantum mechanics postulates a more holistic view of the universe, where the properties of individual particles are not independent but rather part of a larger, interconnected system. This holistic view is in contrast to the classical, reductionist approach, which tries to understand the

BEYOND THE ATOMIC LEVEL

behaviour of the whole by analyzing its individual parts in isolation. Another important implication of quantum mechanics is the observer effect on reality. Classically, properties are held to be independent of observation; however, in quantum mechanics, the act of measurement collapses the wavefunction to determine the state of a particle. This element introduces subjectivity into the process of measurement, which brings out the fact that it is the observer who seems to determine the state of the system.

"Spooky action at a distance" forces a radical reevaluation of the principles that govern our understanding of the universe. The nonlocality that exists in quantum mechanics hints that the universe is more interconnected and complex than it would be under classical physics. The experimental verification of quantum entanglement showed that space, time, and causality as construed classically must be reconsidered in light of quantum theory. A deeper philosophical as well as scientific conflict drove Einstein to discomfort with "spooky action," opposing the localized, deterministic worldview that classical physics had fostered to the probabilistic nature of nonlocal quantum reality. As we delve deeper into the world of quantum mechanical implications, we must live with a reality much stranger and more interdependent than anyone could have conceived within a classical mindset.

The mathematics behind the description of entangled systems shows how quantum particles may correlate in ways that classically seem impossible. This happens at the heart of quantum mechanics, where entanglement is described by a common wavefunction that describes all states of two or more particles as a single collective rather than individually. It goes without saying that this wavefunction is a mathematical object which, in most cases, denotes $|\psi\rangle$. End. For example, if two particles are entangled, their combined state might be written as a superposition of all possible outcomes for both particles, such as:

$|\psi\rangle = a|00\rangle + b|11\rangle$

Here, the notation $|00\rangle$ and $|11\rangle$ symbolizes the quantum states of the two particles, "0" and "1" could be specific properties. Here, "0" may be spin-up, while "1" might denote spin-down. These are the coefficients a and b, which are complex numbers describing the probability amplitude of the corresponding state. When such a system is measured, there will be a collapse to one of the basis states $|00\rangle$ or $|11\rangle$ with probabilities defined by $|a|^2$ and $|b|^2$ respectively.

The key feature of entangled systems is that these states cannot be factored into separate states for each particle. This implies that the properties of each particle are not independent but that the state of one particle is intrinsically connected to the state of the other, regardless of how far apart they are. Mathematically, this is what makes the system entangled. This means the wavefunction will describe the whole system such that measurements on one of the particles instantaneously influences the state of the other, independent of their distances apart.

To study entanglement mathematically, one of the central tools is the correlation function. In quantum mechanics, a correlation function measures the relationship between two observables. For two entangled particles, say, the spin of particle A and the spin of particle B, the correlation function can be written as:

$C(\theta) = \langle A(\theta)B(\theta)\rangle - \langle A(\theta)\rangle\langle B(\theta)\rangle$

Here $A(\theta)$ and $B(\theta)$ are the measurement operators for particles A and B, and $\langle A(\theta)\rangle$ and $\langle B(\theta)\rangle$ are the expected values of these measurements. The angle θ is the angle of orientation of the measuring apparatus. This function allows us to define a way to quantify how strongly the results of the measurements on particle A are correlated with the results of measurements on particle B. These correlations can be very strong in an entangled system, even when the particles are separated by large distances.

If the particles are not entangled, classical physics would predict that the correlation function would yield values consistent with the limits set by Bell's inequalities, derived from the assumption of local hidden variables. However, in quantum systems, the measured correlation function often violates Bell's inequalities, showing that quantum entanglement is a nonlocal phenomenon. That is, the correlations between the measurements on entangled particles cannot be explained by any local hidden variable theory. In other words, the results indicate that entangled particles are connected in a way that violates the classical notions of locality; information about one particle's state is instantaneously communicated to the other, even if separated by vast distances.

To understand the nonlocality mathematically, one can turn to Bell's theorem, which shows that no local hidden variable theory can reproduce the predictions of quantum mechanics. This theorem relies on a mathematical inequality, Bell's inequality, derived assuming locality and realism. Bell's inequality is a condition which any local hidden variable theory should satisfy, but quantum mechanics predicts violations of this inequality. Violations of Bell's inequality constitute an important evidence for quantum nonlocality. For instance, consider the following inequality, a generalization of Bell's inequality for a system of two entangled particles:

$$|C(\theta_1, \theta_2) + C(\theta_1', \theta_2) + C(\theta_1, \theta_2') - C(\theta_1', \theta_2')| \leq 2$$

Where $C(\theta_1, \theta_2)$ is the correlation function between the measurements of two entangled particles at angles θ_1 and θ_2. The violation of this inequality experimentally confirms that the nonlocal correlations are indeed predicted by quantum mechanics, which cannot be explained by any local hidden variables, and the particles are indeed exhibiting "spooky" action at a distance.

Another mathematical model used for entanglement quantization is the density matrix. ϱ is a kind of mathematical representation of any quantum state, especially useful while operating with mixed states and for systems that are connected to an external environment or mixed with it. As this is a two-body problem, the density matrix expression in general is given by

$$\varrho = |\psi\rangle\langle\psi|$$

Where $|\psi\rangle$ is the entangled wavefunction. The purity of the state, that measures how mixed or pure the state is, can be computed as:

$$P = Tr(\varrho^2)$$

Now a state, in general, to have 1 purity and also for a maximally entangled pure state and will be less than one. And it is less as a mixed state where system less entangled or may be statistical mixture of different states. Another important way to express the entanglement of two systems is a given von Neumann entropy from:

$$S(\varrho) = -Tr(\varrho \log \varrho)$$

This entropy measures the amount of entanglement in the system. The higher the von Neumann entropy, the more entangled the state is, while low entropy suggests that particles are weakly entangled or even independent.

The mathematical models give a way to understand the deep connection between the entangled particles and offer concrete ways to quantify entanglement.

They also permit physicists to study the way that these correlations between particles represent some kind of demonstration of quantum mechanics being nonlocal while, at the same time, being the complete opposite for classical physics and locality assumed in its theories with separately independent particle states. Correlations found in entangled quantum models demonstrate how deep in reality and connection classical physics ever could have imagined that really exists.

BEYOND THE ATOMIC LEVEL

Quantum entanglement shows a world of nonseparability, a world where boundaries between things are far more fluid than we have ever been allowed to think under classical paradigms. When these entangled particles are measured, the results show correlations between them that can't be explained by any local process. These correlations are instantaneous in countless experiments, speaking of a depth of connectedness beyond spacetime. This interconnectedness is not limited to a single pair of particles but suggests a holistic framework where all matter may share some level of unity.

Philosophically, this interconnectedness aligns with ancient metaphysical ideas. Many Eastern philosophies, such as Advaita Vedanta in Hinduism and certain Buddhist schools, posit that all phenomena are interconnected aspects of a singular reality. Similarly, indigenous cosmologies often describe the universe as a web of relationships rather than a collection of isolated entities. Quantum entanglement lends scientific weight to these perspectives, providing a modern lens through which to reconsider such timeless ideas. It challenges the materialist worldview that has dominated Western thought since the Enlightenment, calling into question the assumption that reality is fundamentally reducible to separate, independent components.

This is profoundly implicated. If the universe is fundamentally interconnected, the boundary lines we draw, including self and other, and subject and object, all become arbitrary constructs. And it challenges this idea of individuality, at the quantum level, in such a way that this "me" and "not me" distinction dissolves into a seamless fabric. In this sense, quantum entanglement not only offers insights into the workings of the physical universe but also invites us to rethink our place within it. What does it mean to be an individual in a universe that is fundamentally unified?

Are our thoughts, actions, and existence themselves entangled with the fabric of reality?

This shift from separateness to unity is not of the theoretical order; on the contrary, it closely resonates with the unity sought in the unified theory of physics. Scientists over time have searched for a theory that unifies the various forces of nature—gravity, electromagnetism, and the strong and weak nuclear forces—together within a single coherent theory. Quantum mechanics and general relativity, the twin pillars of modern physics, are very successful in their domains but cannot be reconciled at their extremes. Entanglement is a bridge between these theories, which could suggest that the universe's underlying structure is unified.

Some physicists speculate that entanglement might be the key to understanding this unity. In other interpretations, quantum gravity could be emergent, in that the very fabric of spacetime would itself be made of entangled states. The theoretical models like the holographic principle predict that the universe may have a geometry that might arise from quantum entanglement, suggesting that particle entanglement may spread down to the nature of spacetime itself. If true, this would mean the universe is not a collection of separate points in space but a single, unified entity, with entanglement as its fundamental glue. From a philosophical perspective, this unified vision of the universe calls us to reconsider our understanding of reality. It rejects the reductionistic approach that seeks to analyze the most complex phenomena of nature in the simplest language, instead indicating a whole, in which the links between the parts are as natural as the parts themselves are. The universe is more of an organism than of a machine; it presents itself as a dynamic relation of interconnection that is impossible to grasp if torn from its elements.

Such a worldview has far-reaching implications, not only for science but also for ethics and human experience. If all things in the universe are fundamentally interconnected, then a part cannot be healthy when the whole is sick. Indeed, this concept resonates with environmental ethics, whose principles emphasize the interdependency of all life, as well as social philosophies that stress community and responsibility.

In short, quantum entanglement stands for both scientific discovery and a philosophical challenge. It evokes us to go behind the apparent limits of the individual and separateness and to see ourselves in terms of a vast interconnected web. In this process, it bridges the divide between ancient wisdom and modern science, giving us a glance of a reality that at once is profoundly mysterious and deeply unified. The search for a unified theory of physics may ultimately become a search for this connected reality, a quest for understanding the threads that would bind the universe together in ways that transcend our classical knowledge. Perhaps, as we unravel these threads, we will come closer not only to understanding the cosmos but also to understanding ourselves and our place within this cosmic web.

In quantum computing, entanglement plays a role as a valuable resource for processing information in ways that far exceed classical computers for specific applications. Classical computers encode information in bits that represent 0 or 1 states. Quantum computers, in contrast, use quantum bits, or qubits, which exist in superpositions of both 0 and 1 simultaneously. When qubits are entangled, their states are no longer independent; measuring one qubit immediately provides information about the others. This entanglement is what allows quantum computers to perform massively parallel computations. For example, algorithms such as Shor's algorithm for factoring large numbers or Grover's algorithm for searching an unsorted database rely on entanglement to achieve exponential speedup over their classical counterparts. The interconnectivity of qubits in a quantum processor allows for these incredible abilities, making quantum computing an entirely game-changing tool for problems that are intractable on traditional computers.

Quantum cryptography is another area where entanglement proves to be transformative, especially in the form of quantum key distribution (QKD). The most famous QKD protocol, BB84, uses the principles of quantum mechanics to facilitate secure communication. When entangled particles are used in cryptographic protocols, their shared properties ensure that any attempt at eavesdropping disrupts the entanglement and alerts the communicating parties. This unique feature of quantum entanglement makes it possible to create encryption keys that are provably secure against any future computational advancements, including attacks by quantum computers themselves. One area for Entanglement-based QKD is the Ekert protocol, which relies not on computational complexity but the physical laws to ensure key secrecy at its most fundamental level.

In other quantum technologies besides the quantum computer, Entanglement is playing an ever-increasing role, quantum sensing and quantum networks included. Quantum sensing allows enhanced measurement precision through entangled particles reaching over classical limits of sensitivities. For instance, entangled photons enhance the resolution of imaging systems in astronomical or medical diagnostics applications. In quantum networks, entanglement allows for the concept of a "quantum internet," where information can be transmitted with absolute security and minimal loss over long distances. Techniques such as entanglement swapping and quantum repeaters extend the range of entanglement, overcoming the restrictions of direct transmission and opening the way to global quantum communication networks.

The entanglement between quantum particles not only enables superior technological capabilities but also redefines the paradigm through which we process and secure information. By harnessing this deep feature of quantum mechanics, scientists and engineers are building technologies that redefine the limits of what is possible, transform industries, and open up new avenues for innovation. As research continues, the scope of potential applications of quantum entanglement in technology is expanding and underlining its role as a cornerstone of the quantum revolution.

It basically changes our perception of the interrelation between particles and, by extension, all matter. In classical physics, particles are treated as distinct entities with interactions mediated by forces that operate within space and time. However, entanglement is not in the classical paradigm and shows that particles can exhibit correlations that are beyond locality and separability. This deep interconnection hints at a universe in which the distinctions we make between objects are superficial—a world in which relationships, rather than the objects themselves, are fundamental.

Entanglement shows a paradigm in which the distinction between "self" and "other," or between distinct systems, is blurred at a fundamental level. Once particles are entangled, their individual properties are no longer independent but part of a shared quantum state. This interdependence remains intact regardless of distance, hinting toward a deeper, nonlocal fabric underlying the universe. Such a perspective challenges the classical intuition of causality and locality

BEYOND THE ATOMIC LEVEL

governing interactions while inviting us to consider that the whole is greater than the sum of its parts.

The implications of such a shift extend beyond the realm of physics. Philosophically, entanglement resonates with the ideas of unity and interconnectedness that have been presented in many traditions and systems of thought. It defies reductionist approaches where attempts are made to understand the universe by isolating its components, instead it is a holistic view wherein relationships between parts are as important as the parts themselves. This interlinked framework gives a new scope through which to view such questions about the nature of existence, identity, and our place in the universe.

On a practical level, entanglement provides a basis for revolutionary technologies in quantum computing, cryptography, and communication, showing how this apparently abstract concept might have physical, transformative, real applications. These examples show just how the natural relations of particles can be used for tasks that any classical systems could never dream of: hence, emphasizing the value of looking at the world in this relational paradigm.

In the end, quantum entanglement offers more than a glimpse into the mechanics of particles; it provides a new way of thinking about the universe as a profoundly interconnected entity. By reframing our understanding of relationships—not just between particles, but between all elements of matter and energy—entanglement invites us to explore a reality where the lines between the individual and the collective, the local and the universal, are beautifully and irrevocably blurred. This paradigm shift, still in its infancy, promises to redefine not only physics but our greater understanding of the world and our place in it.

BEYOND THE ATOMIC LEVEL

7: DOES TIME REALLY EXIST?

What would be the most precious gift that a man can give to women?

Probably the greatest gift he can offer her is his time. In this world that moves so fast, his undivided attention is a treasure, truly reflecting his care and commitment. Time is at one and the same time an all-powerful force and a fleeting illusion, always sliding through our fingers, but that shapes every moment we feel. It walks ahead without stop but, in passing, it gives meaning to whatever we do. Time is so powerful in healing and in change and transformation yet is endless in the small moments that we share. Its silent passing teaches us what's valuable in the present moment, reminding us how even in those moments we will need, the most profound times often come when least expected.

Time is the quiet rhythm of existence, an invisible thread weaving the fabric of reality. It marks the difference between what was and what is, and yet its essence remains elusive. From the ticking of a clock to the expansion of galaxies, time governs all, yet it is intangible—a ghostly measure that we neither see nor touch, only sense.

For centuries, time has been a silent companion to the unfolding drama of the universe. It has been a cornerstone of physics. In classical physics, time is treated as a universal constant, a steady and absolute backdrop against which events unfold. Sir Isaac Newton, the architect of classical mechanics, envisioned time as an eternal, unchanging river flowing uniformly, independent of the objects and events within it. This view is termed "absolute time" and, indeed provided a simple yet powerful framework of understanding motion, forces and the mechanics of celestial bodies.

In Newton's equations of motion, time was a parameter, an independent variable that allowed us to measure change and predict the future. The ticking of a clock was a representation of this universal rhythm, unperturbed by the intricacies of matter or energy. Classical physics assumed that time was the same everywhere in the universe—an assumption that worked well within the realm of everyday experience.

But, in physics, this classical image collapsed in the 20th century. Einstein's theory of relativity totally overhauled our idea about time by showing that time was relative, not absolute and a four-dimensional continuum in conjunction with space. In effect, time is extensible as well as compressible due to the relative motion of observers or even under the effect of massive objects. This revolutionary insight changed our view of time as dynamic and variable, no longer the universal constant Newton had envisioned.

Quantum mechanics, developed in the early 20th century, posed another challenge to our understanding of time. In the quantum world, time behaves differently. While relativity treats time as part of the fabric of the universe, quantum mechanics relegates it to a mere external parameter. The fundamental equations of quantum mechanics, like the Schrödinger equation, describe how systems evolve over time, but they do not define what time itself is. Time seems to be a background, unquantized and untouched by the uncertainties that govern other quantities in quantum theory.

This apparent inconsistency brings up deep questions: is time independent, as the Newtonians thought, or relational, existing only as a function of the interactions and processes it describes? Do more fundamental principles give rise to time itself, perhaps as some theories suggest?

These questions bring us to the core of a great puzzle. If time is fundamental, then why do quantum mechanics and relativity, the two pillars of modern physics, treat it so differently? And if it is relational or emergent, then how does this relate to our intuitive experience of the passage of time? These are not just philosophical musings but important scientific questions that probe the limits of our knowledge about the universe.

As we continue into the nature of time, we must remember these questions because they will determine not only our view of the cosmos but also our place within it.

In quantum mechanics, time occupies a rather peculiar status compared to all other concepts in the theory. It is not an observable: there is no operator in quantum mechanics corresponding to time. Instead, time is an externally parametrizing quantity a stage for the quantum motion. In sharp contrast to coordinates and momenta, which are both observables and described as operators, time plays a fundamental, but strangely decoupled, role in quantum theory.

This treatment of time raises very deep questions. If quantum mechanics describes the behaviour of all physical systems, why is time outside its formalism? This approach works well in the case of isolated systems which interact with an external clock. However, when we try to describe the universe as a quantum system—a context where there is no external reference for time—the limitations of this framework become evident.

Time, in general, emerged as a parameter from correlations between quantum systems rather than due to any real flow of time. The perspective of the latter has made some physicists to believe that time might not be fundamental but emergent and can come out from deeper structures of reality.

This becomes even more compelling in the search for a theory of quantum gravity, one that unifies quantum mechanics and general relativity. General relativity treats time as an integral part of the fabric of spacetime, inseparable from space itself. Quantum mechanics, by contrast, demotes it to a passive role. The disparity needs to be resolved through a rethink of what time really is.

Quantum gravity theories describe that the spacetime as well as time is not continuous but instead discrete. Spacetime in loop quantum gravity would hence be made of discrete portions of finite size as that the energy is quantized in quantum mechanics. In this context, therefore, we can imagine that what seems like continuous time flow is derived from just a sequence of discrete ones.

Other theories, such as the Wheeler-DeWitt equation in quantum cosmology, attack the very fabric of what time is at a fundamental level. The Wheeler-DeWitt equation represents the entire universe without a parameter for time. This has implications that the universe is ultimately timeless at its base. Perhaps it is only emergent and born from the correlation of patterns between different parts of space.

This view of time as emergent is not merely a theoretical abstraction. It fits well with some interpretations of quantum mechanics, where the concept of a "clock" becomes increasingly tenuous at microscopic scales. The more we probe the quantum world, the more time as a continuous and universal concept starts to blur into a more relational understanding.

Such ideas challenge the intuitive understanding of time, as an ever-forward-moving river. If time is emergent, then its flow is perhaps an illusion, that is, a large-scale phenomenon arising from interactions at much smaller, timeless scales.

This realization forces us to reconsider the nature of causality, change, and even our place in the universe.

Perhaps, one of the greatest enigmas in physics still revolves around time, which lingers somewhere between the outer limit of quantum mechanics and the inner lining of relativity. Whether this is something to be created or some aftermath, whether a mere effect or some inherent ingredient, the enigma still surrounds time, hiding its deep truth from understanding. By investigating the very nature of time, we approach step by step into discovering the intricate relationship between quantum and cosmos.

In Einstein's theory of general relativity, time is interwoven into the fabric of the universe. Space and time are no longer distinct but are united in a four-dimensional continuum called spacetime. This revolutionary idea transformed our understanding of reality, allowing us to describe how massive objects warp spacetime, giving rise to the effects of gravity. Time in this framework is no longer absolute, as it was once imagined by Newton, but relative; time stretches, compresses, and bends with the observers' motion and with massive objects.

The general relativistic treatment of time is deeply geometric. The events that make up the universe are laid out as points in space-time, and their relation to one another is established through the geometry of the continuum. It also illustrates gravitational time dilation: the clock that is near the surface of a supermassive star ticks more slowly than one that is farther away from such a gravitational field. This interplay of mass, energy, and structure of spacetime demonstrates how the fabric of space and time is not absolute but something like putty.

In contrast, quantum mechanics deals with time in a fundamentally different manner. Here, time does not belong to the system itself but rather is taken as an external parameter whose dynamics do not influence that of the quantum system itself. Relativity drenches time in the system, whereas quantum mechanics keeps it at arm's length, utilizing it to describe the temporal evolution of a wave function without ever asking it to consider its nature.

This basic difference causes a strong tension between the two theories. Time in general relativity is dynamic and is variable according to the system state. Time in quantum mechanics is static, forming a sort of an immoveable backdrop against which the probabilistic dance of particles and wave functions takes place. It makes the two theories not so easy to harmonize into a single frame of unity.

The contradictions become particularly stark when considering scenarios such as black holes or the very early universe, where both quantum effects and the curvature of spacetime are significant. General relativity would suggest that spacetime itself becomes highly distorted, while quantum mechanics requires a stable, external clock to function. How can these two conflicting treatments of time coexist?

Attempts to bridge this gap, as in quantum gravity theories, usually require a reconsideration of time itself. Some theories suggest that time, as understood in classical physics, is not fundamental but emergent—a macroscopic phenomenon arising from the underlying quantum reality. These ideas propose that at the deepest levels of the universe, time might not exist as we know it, and its apparent flow could be a large-scale illusion.

The biggest challenge in modern physics is how to reconcile geometric time as dictated by general relativity with the external parameter used in quantum mechanics. Time, in fact, might be a far more radical concept than any of these, and unifying them into a coherent reality picture would demand a shift in paradigms. For now, time remains one of the greatest enigmas. It is suspended between the relativistic cosmos and the quantum realm.

BEYOND THE ATOMIC LEVEL

Time dilation is one of the fundamental components of Einstein's theory of relativity, showing the elastic nature of time in the universe. Time does not flow uniformly for all observers; its passage depends on factors like relative velocity and the presence of gravitational fields. For example, a clock moving at high speeds or situated in a strong gravitational field ticks slower compared to one at rest or farther from massive objects. This effect, as confirmed by atomic clocks and GPS satellites experiments, puts forward the view that time is not a constant but rather a relative experience bound to spacetime. In contrast, quantum mechanics introduces the concept of timelessness at a fundamental level.

Unlike relativity where time is a dynamic dimension of spacetime, quantum mechanics treats time as an external, unchanging parameter. The quantum systems evolve in accordance with the Schrödinger equation, which involves a universal time parameter unaffected by the quantum system. Time in this framework serves as a marker for the progression of a wavefunction but is not a physical quantity intrinsic to the system.

The divergence becomes clear when one considers the entire universe. In quantum cosmology, equations such as the Wheeler-DeWitt equation describe the universe in a "timeless" state. In this state, no explicit role is given for time as an evolving parameter; instead, time could emerge as a macroscopic phenomenon from correlations between parts of the system. This concept is known as emergent time, where seemingly flowing time might be some emergent phenomenon out of the more fundamental timeless reality of the quantum world. In short, thus the juxtaposition of time dilation with timeless quantum systems highlights one of the deepest mysteries ever. Relativity insists upon the relativity and variability of time; quantum mechanics puts it on an exterior framework or emergent in some way. This remains one of the greatest challenges of modern physics: it says

BEYOND THE ATOMIC LEVEL

something deeper and unified to understand the universe.

Imagine time as a river, and every person, object, is a floating boat.

To most people, the river seems fluid and constant. However, depending on where you position yourself on the river, the current changes. Suppose you are close to where the river slows down because obstacles have slowed it down somewhat; your boat moves rather slowly. Now if you happen to be in the middle, where the current flows strongly, then your boat increases speed. Imagine now this river represents time passing.

The "current" of time now flows according to where or how fast you are situated. The closer you get to some huge mass object, such as a planet or star (the riverbank), the slower the passing of time for you is. If you are moving fast as if you were floating down the middle of the river, time seems to pass more quickly. To someone standing far off, untouched by the passage of time, your own experience of time may differ from that. In this analogy, just as the river's current is influenced by where you are and how you move, time behaves similarly—it can stretch, slow down, or speed up depending on your position and motion in the universe. But no matter what, everyone experiences time relative to their own perspective, even if that experience is not the same for everyone else.

The Wheeler-DeWitt equation is $\hat{H}\Psi = 0$. It is the core idea of quantum cosmology: it describes the quantum state of the whole universe. Here, $\hat{H}$ is the Hamiltonian operator (which is essentially the total energy), and Ψ is the wave function of the universe. This formulation does not contain time as an explicit parameter, unlike the Schrödinger equation.

This "timelessness" is perhaps one of the most interesting features of the equation. It makes the universe appear, at its most basic level, to be an eternal quantum state, a static object. What we experience as the passage of time might only arise as an effect when particular subsystems of the universe are measured. Thus, for example, the correlations between various parts of the wave function could make it seem that something has happened over time—what we experience as the flow of time.

The timeless nature of the Wheeler-DeWitt equation challenges our understanding of reality. It implies that time, rather than being a fundamental aspect of the universe, is a construct that arises from our perspective as observers. This aligns with the idea of emergent time, where the passage of time is a large-scale phenomenon stemming from the interactions of smaller, more fundamental elements.

While these implications are profound, they remain speculative, as a complete theory of quantum gravity is still out of reach. However, the Wheeler-DeWitt equation offers a tantalizing glimpse into a universe where time, as we know it, may not exist at the most fundamental level.

All of us experience time as flowing from the past into the future, but why it is so and not the reverse is a profound challenge that lies at the heart of philosophical and scientific inquiry.

One of the most widely accepted explanations for the arrow of time is the second law of thermodynamics, where it is stated that for any isolated system, total entropy always increases with time. This increase in entropy not only describes the direction of time, but it also offers one explanation as to why things move one way but can never retrograde.

The divergence in our understanding of time becomes stark when we examine the universe as a whole, particularly in the context of quantum cosmology. Equations like the Wheeler-DeWitt equation depict the universe in a "timeless" state, where time does not explicitly appear as an evolving parameter. Instead, time may emerge as a macroscopic phenomenon, arising from the correlations between different parts of a fundamentally timeless quantum system. This concept, known as emergent time, proposes that what we perceive as the flow of time is the result of complex interactions within a timeless quantum reality.

This notion underscores the profound difference between relativity and quantum mechanics. In relativity, time is relative and malleable, intricately woven with space into a dynamic spacetime fabric. It stretches, compresses, and varies depending on the observer's motion or gravitational influences. Conversely, quantum mechanics treats time as an external parameter or an emergent property, not a fundamental aspect of the universe. Bridging these divergent views remains one of the most significant challenges in modern physics, hinting at a deeper, unified framework.

To imagine this, consider time as a river. Each object or individual is a boat floating on it. For most observers, the river flows smoothly and at a consistent pace. Yet, depending on the boat's position or speed, the experience changes. Near the riverbank, where obstacles slow the current, the boat moves slower. In the middle, where the current is stronger, the boat accelerates. Similarly, the flow of time varies based on one's proximity to massive objects (gravity) or their relative velocity.

While each person experiences their own "current," the flow of time remains relative, differing from one observer to another but remaining valid within their perspective.

At the heart of quantum cosmology lies the Wheeler-DeWitt equation, represented mathematically as $\hat{H}\Psi = 0$. Here, $\hat{H}$ denotes the Hamiltonian operator (representing the universe's total energy), and Ψ is the wave function describing the quantum state of the universe. Unlike the Schrödinger equation, this formulation omits time as an explicit parameter, suggesting a universe that exists in a static quantum state, unchanging and eternal.

BEYOND THE ATOMIC LEVEL

Our perception of time's flow might emerge only when observing specific subsystems of the universe. Interactions and correlations within the wave function give rise to the apparent progression of events—what we recognize as time. This timeless nature challenges deeply held notions of reality, proposing that time may not be a fundamental component of the universe but rather a construct derived from our observational perspective.

The idea of emergent time aligns with another profound concept—the arrow of time. Why does time seem to flow in one direction, from the past to the future? The second law of thermodynamics offers a compelling explanation: in any isolated system, entropy (a measure of disorder) increases over time. This increase establishes the forward direction of time and explains why time seems irreversible in our experience.

While the Wheeler-DeWitt equation and the idea of emergent time provide tantalizing insights, they remain speculative. A complete theory of quantum gravity, which unites relativity and quantum mechanics, is necessary to fully grasp the nature of time. Nonetheless, these concepts hint at a universe where time as we know it may not exist at the most fundamental level, reshaping our understanding of existence and the cosmos.

This concept uses the explanation of entropy, or the measure of the disorder or randomness within a system. In simple words, entropy is described as the number of possible configurations or states a system can assume. The second law of thermodynamics dictates that in the course of time, natural processes will increase the total entropy of a system. This increase is irreversible, meaning that once the system has moved toward a higher-entropy state, it does not spontaneously return to a lower-entropy state without external intervention.

This increase in entropy is the key to understanding the

arrow of time. The idea is that the progression of time is closely tied to the movement from lower-entropy states to higher-entropy states. The past can be thought of as the time when the universe was low in entropy and the future as the time when entropy has been increased. It is this difference between the low-entropy past and the high-entropy future that gives the "direction" of time-the arrow that points from the past to the future.

A common example that illustrates the second law of thermodynamics is the process of mixing hot and cold water. If you take a glass of hot water and pour it into a glass of cold water, the heat will distribute itself throughout the water, and then the system will be equilibrated to a uniform temperature. This is an irreversible process: the heat will not spontaneously collect itself back into the hot water without some external force or intervention. The entropy of the system increases as the heat disperses. This increase in entropy gives the forward-moving arrow of time, as the process can only go in one direction—that is, towards greater uniformity and disorder.

Another example can be seen with a broken vase. When a vase is broken, its fragments are distributed in nonordered positions. The action of breaking the vase increases the entropy of the system because the pieces move from a low-entropy, ordered condition (a whole vase) to a higher-entropy, nonordered condition (shattered pieces). The reverse process in which the pieces spontaneously organize into a vase does not happen naturally. This irreversible transformation again shows how the second law of thermodynamics propels the arrow of time forward, from a more ordered state to a more disordered one.

In a cosmological context, the second law of thermodynamics has even deeper implications. The universe, at the time of the Big Bang, was in a state of extremely low entropy. Over time, with the expansion of the universe and its evolution into what it is today, entropy has increased. This means, in the current state of affairs, systems are in more disordered states compared to their previous states. In a way, the future will eventually reach what is called "heat death," where maximized entropy is reached as all energy is evenly dispersed, allowing no possibility to do any work or effect any change. The increasing entropy of the universe, therefore, defines its future trajectory and establishes a cosmic arrow of time.

This irreversible increase in entropy is what gives time its one-way flow. It explains why we can remember the past but not the future, and why natural processes tend to move toward greater disorder. The second law of thermodynamics gives us a scientific framework for understanding the arrow of time-not as some human construct, but as a fundamental feature of the universe, based on the natural tendency of systems to evolve from order to disorder. Thus, time, in its nature, is tied to the entropy increase, always moving from the low-entropy past to the high-entropy future, and making the direction of time one of the most fascinating aspects of the universe.

In classical thermodynamics, the difference between reversible and irreversible processes is well-defined. A reversible process is one that can be reversed without leaving any change, while an irreversible process is a one-way transformation toward greater disorder or entropy. This distinction is influenced by the second law of thermodynamics, which governs the natural tendency for entropy to increase in closed systems. However, when we delve into the quantum realm, the concepts of reversibility and irreversibility become more nuanced due to the peculiar behaviour of quantum systems.

At the heart of quantum mechanics lies the idea that systems evolve over time according to the principles of wave functions, superposition, and probability. Quantum systems can experience both reversible and irreversible processes, but the conditions under which these occur differ significantly from classical systems. To understand this, we need to explore how these processes manifest in the quantum world.

In quantum mechanics, reversible processes are usually described by unitary transformations. A unitary operation is a mathematical transformation that preserves the total probability—or the total "information"—of a quantum system. The evolution of a quantum state under a unitary operator is time-reversible. This means that if you reverse the process, you can return the system to its original state without any loss of information. In quantum mechanics, this is the characteristic that underlies reversibility in quantum systems.

One of the most important examples of a reversible quantum process is the time-evolution of a closed quantum system described by the Schrödinger equation. This equation describes the unitary evolution of the wave function that describes the state of a quantum system. This means that if we know the wave function at one point in time, we can theoretically calculate its state at any other time, both forward and backward. This time-reversible behaviour is a defining feature of quantum systems.

For example, consider a free quantum particle in a potential. Its evolution in time is given by the Schrödinger equation, describing an unitary process. Hence, the system may move from one state to any other, and vice versa. If the wave function is known at one time it is possible to predict it everywhere else, and the evolution is reversible.

While unitary evolution describes quantum mechanics in ideal conditions, there are processes within the quantum realm that appear to be irreversible. Irreversibility is most commonly associated with quantum measurement and decoherence. Quantum mechanics does not describe particles having definite properties until they are measured. Instead, they exist in a superposition of all possible states.

When an observation is made, this superposition collapses into a single definite state, and the wave function undergoes what is called a "collapse." This collapse is non-unitary and represents an irreversible change, as the system loses the information about its previous superposition of states.

This process is closely related to quantum decoherence, which occurs when a quantum system interacts with its environment. Decoherence causes the quantum system to lose its coherent quantum properties and behave in a classical manner. Once a quantum system becomes entangled with its environment, it effectively "forgets" its quantum nature and is forced into a mixed state. The system no longer behaves in a purely quantum way and its evolution becomes effectively irreversible.

For instance, when a qubit represents information in a quantum computer by existing in a superposition of states, then through noise, thermal fluctuations, or measurements, its decoherence with the environment tends to lose its quantum coherence to result in a superposition collapsing into a classical state. Decoherence is in a way irreversible since once quantum information has been dispersed to an environment, it cannot be recalled without external intervention.

In classical thermodynamics, irreversibility is closely related to the second law, which states that entropy is always increasing in an isolated system. However, in the quantum regime, the second law is not as simple. Quantum systems can be in superpositions of states that allow for both reversible and irreversible behaviours. The irreversible aspects of quantum systems are mainly seen when measurement or decoherence occurs, which causes a loss of quantum coherence and a transition to classical behaviour.

Despite the reversibility inherent in quantum mechanics under ideal conditions, the interaction of quantum systems with their environments and the process of measurement give rise to irreversible changes. Entropy in quantum mechanics is described using von Neumann entropy, which measures the degree of "mixedness" or disorder within a quantum system. When a quantum system interacts with its environment, the system becomes entangled, and its state transitions from a pure state (low entropy) to a mixed state (high entropy). This process is irreversible, similar to the classical idea of increasing entropy.

The arrow of time in quantum mechanics, much like in classical systems, is linked to the increase in entropy. The forward progression of time corresponds to the system's transition from low-entropy to high-entropy states. In quantum mechanics, this transition is closely tied to the measurement or decoherence of a quantum system. When a quantum system interacts with its environment, quantum coherence is lost, and the system evolves toward classical states. This process introduces an irreversible change that aligns with our perception of the "one-way" flow of time.

The evolution of a quantum system over time is governed by the time-evolution operator, which encapsulates the dynamics of the system. Represented as U(t), the time-evolution operator describes how a quantum state evolves from its initial configuration at t = 0 to a state at a later time t. The mathematical expression for this operator is:

$$U(t) = e^{-iHt/\hbar}$$

The time-evolution operator provides a complete description of how quantum states change over time. If a quantum system begins in the state $|\psi(0)\rangle$ at t = 0, the state at a later time t is obtained by applying the operator U(t) to the initial state:

$$|\psi(t)\rangle = U(t)|\psi(0)\rangle$$

This equation illustrates that the evolution of the quantum state is determined entirely by the time-evolution operator, which itself depends on the system's Hamiltonian. The Hamiltonian encodes the dynamics of the system, including its energy and interactions, making it a fundamental element in understanding the behaviour of quantum systems.

A crucial property of the time-evolution operator is its unitarity. This ensures that the operator preserves the total probability in the system over time, which is a cornerstone of quantum mechanics. Mathematically, unitarity is expressed as:

$$U^\dagger(t)U(t) = U(t)U^\dagger(t) = I$$

where U†(t) is the Hermitian conjugate of U(t), and I is the identity operator. The unitarity of U(t) guarantees that if the quantum system starts in a normalized state, the state remains normalized throughout its evolution. This preservation of normalization ensures that probabilities derived from the quantum state remain consistent and physically meaningful, reflecting the core principles of quantum theory.

For a time-independent Hamiltonian, the evolution of a quantum system is relatively straightforward. The time-evolution operator takes the form of a simple exponential function, directly facilitating the calculation of the quantum state's progression over time. However, when the Hamiltonian is time-dependent—indicating that the system's energy or interactions change over time—the time-evolution operator becomes significantly more intricate. In such cases, the evolution is described by a time-ordered exponential, which accounts for the system's dynamic changes.

The time-evolution operator is fundamentally linked to the Schrödinger equation, which governs the temporal behaviour of quantum systems. The time-dependent Schrödinger equation is expressed as:

$$i\hbar \frac{\partial}{\partial t} |\psi(t)\rangle = H |\psi(t)\rangle$$

The solution to this equation is provided by the time-evolution operator. If the system's initial state $|\psi(0)\rangle|$ at $t = 0$ is known, the state at a later time t can be calculated using

the operator:
$$|\psi(t)\rangle = e^{-iHt/\hbar} |\psi(0)\rangle$$
This formulation illustrates how the time-evolution operator encapsulates the system's dynamics and enables predictions about its future behaviour.

For systems with a time-independent Hamiltonian, the evolution of the wave function is governed by a constant exponential factor, making the calculation straightforward. However, for systems where the Hamiltonian changes with time—such as those subjected to external forces or varying potentials—the time-evolution operator requires a time-ordered exponential to reflect these dynamic conditions. This added complexity is necessary to accurately describe how the system's interactions and energy change over time.

Consider, for instance, a quantum particle moving in a potential. The Hamiltonian for this system includes terms representing both the particle's kinetic energy and its potential energy. If the potential remains constant, the time-evolution operator is simple and straightforward. However, if the potential varies with time, the time-evolution operator must adapt to these changes, leading to more complicated dynamics. This reflects the necessity of the time-ordered exponential in capturing the system's behaviour accurately.

The time-evolution operator, whether in its simple exponential form or the more complex time-ordered exponential, is a cornerstone of quantum mechanics. Its unitary nature ensures the conservation of probability, maintaining the normalization of the wave function throughout the system's evolution. By encapsulating the effects of the Hamiltonian—be it time-independent or time-dependent—the time-evolution operator provides a robust framework for understanding the dynamics of quantum systems. It offers profound insights into how these systems evolve over time, shedding light on the fundamental workings of the universe.

BEYOND THE ATOMIC LEVEL

Standard time-evolution in traditional quantum mechanics usually runs from the past and forward into the future. On the other hand, time-symmetric approaches assume that quantum equations may equally be valid if run backward and forward in time.

Conventional quantum mechanics usually bases the evolution of a system on the Schrödinger equation. Its state at any time depends on the states at previous times. This one-way flow is therefore a reflection of our intuitive experience in everyday life with time. Formulations that are time-symmetric, however, suggest the laws of quantum mechanics can't favor a particular orientation of time, which may reflect that the future is as determinable by the past as is the past by the future.

A key concept in time-symmetric quantum mechanics is that the dynamics of quantum systems could be formulated in such a way that the same mathematical framework describes the system's behaviour both forwards and backwards in time. This is quite different from the standard interpretation, where time is treated as an arrow with an irreversible flow from past to future.

One of the most significant time-symmetric interpretations is the retrocausal interpretation of quantum mechanics. It postulates that in a certain way, future events can influence past events. It further means that measurements carried out in the future can influence the result of experiments performed in the past. This contradicts the traditional view of causality. In this view, causes precede effects. Retrocausality does not refer to the journey from future to past as that term is typically conceived in time travel. The borderline between past and future should not be viewed so precisely, as one has long come to understand it.

Outcomes of quantum experiments are determined not by how the system is described by the state at the present, but by possible measurements to be taken in the future

An additional example of time-symmetry occurs within Feynman's path-integral formulation. In this approach, the probability amplitude for a particle going from one point to another, it is not only summed over all possible paths it may take forward in time, but also over all of the possible paths it might have taken backward in time. This means that this formulation is symmetric under the exchange of past and future, which allows for a more holistic approach to quantum processes.

In addition, the time-symmetric perspective highlights advanced and retarded potentials in quantum field theory. In this point of view, fields as well as particles are controlled by advanced solutions as well as the past ones, which are the retarded solutions, and it suggests a possibility of time-symmetric interaction between particles and fields.

Of course, time-symmetric quantum mechanics poses many questions into how we perceive and then interact with time. Time runs only one way in our experience, namely forward, from past through to present and into future. This directionality of time is reflected in the second law of thermodynamics: that entropy, or disorder, always tends to increase with time. This asymmetry between past and future is in complete contradiction with time-symmetric formulations of quantum mechanics, which are indifferent to time direction. The reconciliation of these two views—time-symmetric quantum mechanics and the arrow of time—remains one of the open questions in physics.

Moreover, time-symmetric approaches face the measurement problem in quantum mechanics, involving the role of the observer and the collapse of the wave function. If quantum mechanics is time-symmetric, then it could be that the measurement process is equally determined by both past and future factors. This would go against the standard interpretation, where the measurement process is viewed as an essentially irreversible event that collapses the wave function and determines a specific outcome.

Imagine you are trying to measure the speed of a car on a curving, winding road. You do so by tracking the car's position at regular intervals and determining the distance it traveled between them. The more often you measure the position of the car, the better will be your measurement of distance covered and therefore speed. But as the intervals get small, the problem increases: speed fluctuations in the car go up because of the curve in the road, and measurements become increasingly uncertain.

Replace the car with a quantum system such as an atom, and position measurements by energy transitions. Just like the speed of the car fluctuates, the quantum system's energy levels oscillate. Quantum clock ticks through measuring oscillations: a very high resolution "ticking" of time. But just as the fluctuations in the car's speed have limits, so does the quantum system's measurements. The more accurately we attempt to measure the quantum system's "speed" (or energy), the more imprecise our "timing" becomes as a result of the intrinsic limitations that quantum mechanics enforces, such as the time-energy uncertainty relation. In exactly the same way that the car's speed cannot be measured with arbitrary precision on the curved road, the quantum clock cannot be kept perfectly in step.

These clocks operate based on the principles of quantum systems, and their basic nature specifically: oscillatory motion. There is, obviously, a most celebrated model for timekeeping based upon the vibrational behaviour of atoms ; such models define the quantum systems involved in oscillatory behaviour relative to atomic vibrations in quantum systems, wherein those very oscillations translate directly to the atomic states involved with energy and accurately yield measurement of time itself.

At its very core lies the energy level of a quantum clock which is usually within an atom or any system. While an atom absorbs or releases electromagnetic radiation, it would change energy states in small discrete increments. These energy transitions come with timescale precision; hence quantum clocks can provide time measurement accurately. The frequency of these transitions can be considered as a "quantum standard" for timekeeping, which has led to the development of highly accurate atomic clocks that form the backbone of modern timekeeping systems like GPS.

Quantum clocks face significant challenges, one of the most prominent being the time-energy uncertainty relation in quantum mechanics. This fundamental principle is expressed mathematically as:

$$\Delta E \Delta t \geq \frac{\hbar}{2}$$

Here, ΔE represents the uncertainty in energy, Δt represents the uncertainty in time, and $\hbar$ is the reduced Planck constant. This relation imposes a fundamental limit on the simultaneous precision of energy and time measurements.

The time-energy uncertainty relation reveals an inherent trade-off: the more precisely we attempt to measure the energy of a quantum system, the less precisely we can measure the corresponding time interval, and vice versa. This limitation is not due to technical shortcomings but arises from the very nature of quantum mechanics. It reflects the fundamental quantum behaviour of systems, where certain pairs of quantities—such as energy and time—cannot both be precisely determined simultaneously.

For quantum clocks, this poses a significant challenge. While they aim to measure time with exceptional precision, doing so may increase the uncertainty in the energy of the system being measured. This constraint underscores the delicate balance required in designing and utilizing quantum clocks, as well as the deeper mysteries of time and measurement in the quantum realm.

BEYOND THE ATOMIC LEVEL

This would imply a limitation on the quantum definition of time. The uncertainty principle shows that, in order to measure time with a high precision, we have to tolerate some uncertainty in the energy of the system. This becomes an important trade-off when one tries to create clocks operating at extremely small time scales where quantum effects become more prominent.

Even when defining time, practical difficulties arise. It cannot be said that in the case of classical physics, time is easily measurable and treated as a continuum. Quantum mechanics complicates this picture by introducing time as something that could not possibly be so discrete. Because quantum systems behave in terms of superposition and entanglement, there seems to be no definite time at play here either.

More importantly, quantum systems are not easily measured with a ticking clock. Indeed, in most experiments, decoherence would ruin the measurements if the quantum system is exposed to external influences. At such small scales, environmental noise is significant and could interfere with the delicate quantum states that are necessary for the proper timekeeping.

Quantum clocks play an essential role in modern physics experiments, including those in quantum information science and general relativity. For example, precision measurements of time are critical for the testing of the fundamental theories of physics, such as possible variations in the fundamental constants of nature with time. Quantum clocks have also been used in experiments to check how gravity affects time. It helps understand the relationship between time and space as predicted by Einstein's theory of relativity.

Practical application of quantum clocks can be envisioned in applications where precision timing becomes a core component of that technology, such as in the case of GPS and telecommunication. High accuracy achieved with quantum clocks can facilitate synchrony across distances so that huge systems are precisely coordinated with each other for such technology to function properly.

Emergent time theories postulate that, rather than being fundamental aspects of reality, time appears as a macroscopic phenomenon created from more fundamental and time-less processes. This view challenges the classical and quantum mechanical understanding of time as an intrinsic feature of the universe, suggesting that time is a product of some underlying physical laws that can only be seen at larger, observable scales. In such theories, time is not a basic building block of the universe but a consequence of more fundamental phenomena that govern the microcosm of quantum systems.

One of the key areas where this idea gains traction is in the realm of spacetime foam and Planck scale physics. At the Planck scale, which corresponds to incredibly small distances—around 10^{-35} meters—spacetime itself is believed to be highly turbulent and chaotic. This "spacetime foam" is not smooth and continuous, but rather a frothy, fluctuating structure where traditional notions like space and time lose their usual meaning. At such scales, the notion of time as a continuous, linear progression breaks down, and it is speculated that time, like space, might emerge as a large-scale feature of more fundamental quantum events.

Loop quantum gravity is one of the most prevalent theories in quantum gravity, through which the concept of emergent time becomes more understandable. In this approach, space and time are quantized down to the smallest scales, the fabric of spacetime having discrete granular structures composed of it. Time emerges from the interactions and relationships in this context between these discrete structures. The theory suggests that, at the Planck scale, time is "frozen," in a sense, and only when viewed at larger scales does the appearance of continuous time arise.

BEYOND THE ATOMIC LEVEL

This emergent time perspective implies that the flow of time we experience is a macroscopic phenomenon emerging from the quantum interactions of spacetime at a level too small to directly observe.

Similarly, string theory gives hints of emergent time, but it frames this concept in a slightly different manner. In string theory, matter is not made up of point-like particles but rather by one-dimensional "strings" vibrating at different frequencies. These strings exist in higher-dimensional space, and their interaction gives rise to the familiar forces and particles in our universe. String theory posits that time could emerge from the way these strings interact and vibrate in higher dimensions. Just as spacetime in general relativity is a smooth continuum, in string theory, time could be a macroscopic phenomenon that emerges from more fundamental interactions within these higher-dimensional spaces. The vibrational pattern of strings and their interaction with one another may lead to flow, which we understand the concept of time to be inside this universe. Time is certainly not an inherent characteristic of strings; it's an aspect emerging from the larger dynamical scales.

This will further suggest that perhaps our experienced time is not fundamental; perhaps it's emergent too. Instead, it may arise from deeper, more fundamental structures that are not accessible to our observations. In the microscopic world of quantum mechanics, time behaves very differently, and it is only at macroscopic scales, when quantum effects average out, that we experience the time we are familiar with. As our insights into quantum gravity, spacetime, and the universe go deeper, theories such as loop quantum gravity and string theory may more clearly describe just how time arises from the workings of the more fundamental forces and particles, reshaping our comprehension of this deeply imbedded aspect of reality.

The nature of time plays a crucial role in shaping

cosmological models, influencing how we understand the origins, evolution, and ultimate fate of the universe. For centuries, time was thought to be a simple, linear progression—an ever-present backdrop against which events unfold. However, as our understanding of the universe has progressed, especially with the development of quantum mechanics and general relativity, the nature of time becomes more complicated and entangled in the fabric of spacetime itself. This shift in our thought has important implications for cosmology and for developing a unified theory of quantum gravity.

In the general relativity framework, time is a dimension that is part of the fabric of spacetime. Time is relative; it depends on the observer's motion and gravitational field. This has dramatically changed our understanding of the cosmos, as time can even bend with the curvature of spacetime. For example, close to a very heavy object, such as a black hole, time goes by more slowly than elsewhere for an observer far away; this is called gravitational time dilation. These phenomena are not just theoretical but were experimentally confirmed, like the observation of atomic clocks in satellites orbiting the Earth, where time flows slightly faster than on Earth's surface due to different gravity.

Even as such the nature of time is so bewildering when looked at a much smaller scale than the cosmic. Quantum mechanics, while governing the behaviour of energy and particles at the small scales, is much more different from relativity and treats time very differently too. In quantum theory time is usually treated as the external parameter which is under no uncertainty or warping like that governing space in relativity. This discrepancy between the way time has been treated in quantum mechanics versus general relativity has led to one of the most important challenges of modern physics, that is, the quest for a theory of quantum gravity.

Quantum gravity seeks to reconcile these two apparently

incompatible theories, that is, quantum mechanics and general relativity, to give an account of the universe on all scales with a single framework. In this search, the nature of time is a central issue. Certain approaches, such as loop quantum gravity, indicate that time, as we are familiar with it, is not fundamental at all but rather may emerge from quantum interactions of spacetime at extremely small scales. In this way, time is not as it is at the quantum level as it is in our everyday experience. Time may not be a continuous flow but rather a macroscopic phenomenon that emerges only when viewed from the larger scales of our universe.

String theory is another leading candidate for a theory of quantum gravity. String theory gives a different view. In string theory, time may be an emergent property that arises from the interactions of fundamental strings in a higher-dimensional space. Just as the familiar three dimensions of space are a consequence of higher-dimensional geometry, time might be a consequence of the dynamics of the strings in these extra dimensions. Both approaches imply that time's nature is not predetermined but rather a product of the underlying quantum structure of the universe.

The effects on our understanding of time due to these new theories are profound. If time is indeed an emergent property, then this would call into question our traditional concepts of causality and the progression of events. It means that perhaps time, as one of the fundamental properties of the universe, does not exist in the way we are experiencing it. Instead, it may be a byproduct of the underlying laws of quantum mechanics and gravity, which only manifest themselves at the macroscopic level. This would make a complete rethinking of the thought process regarding the origin of the universe, its form, and its future.

In addition, these theories also affect our perception about the start and end of the universe. In the standard cosmological model, time starts at the Big Bang, a singularity where the laws of physics break down. If time is emergent, then the notion of a beginning point loses meaning and the universe may have no true origin, but a transition from a timeless state to the emergence of events as we understand them. Similarly, if time is not fundamental, the end of the universe need not be a "final moment" but a continuation of processes beyond our present comprehension.

In conclusion, the nature of time has profound implications for cosmological models and our understanding of the universe itself. The search for a theory of quantum gravity not only promises to unify the fundamental forces of nature but also forces us to reconsider our most basic assumptions about time. As our understanding deepens, we may find that time is not a constant, immutable force but a product of deeper quantum realities, reshaping how we perceive the universe's origins, structure, and ultimate fate.

The nature of time has long been an important subject of profound philosophical debate, with two prevailing positions being in opposition to each other: time as something abstract and time as something essential. These two perspectives shape most of our understanding of the universe and guide scientific inquiry, while inspiring metaphysical reflection.

From one perspective, time appears as an abstract concept—a creation of human beings to mark the change and order in life. In this, it is not something that comes out of the observer itself, but rather a method that allows us to rationalize the sequence of happenings. This constructivist view of time is that it is a mental framework, dependent on our perception and not an inherent part of the world itself. Time, in this regard, is molded by the way we think about and organize our experiences rather than it being a fundamental, physical entity.

There is also the view that time is an intrinsic part of reality—a fundamental and objective feature of the universe. In this perspective, time exists independently of human perception and serves as the central element of the laws of physics. Time, in classical mechanics, is absolute, an uninterrupted flow in which all events occur as a backdrop to them; it is that against which all motion is measured, and it controls the course of the universe. This is further supported by the fact that Newtonian physics was so successful and intuitively appealing: time is a river, flowing relentlessly forward, and nothing can affect its passage by our observations or experiences.

But the deeper our comprehension of the universe, the more it becomes mysterious. Through quantum mechanics and general relativity, time has turned out to be a mysterious variable: in quantum mechanics, it plays differently; it is an external parameter and does not relate to the quantum states of particles. Thus, this view challenges the way time has been classically conceived as a flowing entity of its own, suggesting instead that time, in this quantum world, is actually lesser and more fluid. Indeed, general relativity has pointed to the fact that time has no absolute nature but warps in response to the actions of gravity and motion. This relativistic view of time complicates the idea of an invariant, universal clock and instead proposes that time is intertwined with the spacetime fabric, being affected by the existence of mass and energy.

These two conflicting theories raise basic open questions regarding the true nature of time. One of the most interesting questions is whether time can be quantized. In the quantum world, physical quantities like energy and momentum are quantized; they come in specific, quantized amounts. Could time be quantized too? Might it be that at the Planck scale, time appears in small, indivisible units? If so, this could challenge our intuitive understanding of continuous time and reshape our views of events, causality, and the flow of existence itself.

Another urgent question concerns the objective existence of the flow of time. In our everyday experience, time seems to pass irreversibly. It flows from past to future in a constant stream, but is this flow an objective feature of the universe, or merely a product of our perception? Does the passage of time have independent existence outside the way we experience it? These are questions that have far-reaching implications, not only for our understanding of time but for the very structure of reality. If the flow of time is subjective, then this would mean that the universe itself does not have a real, objective "flow of time" but rather this flow is a consequence of our particular perspective on the cosmos.

These open questions about time continue to be central to the search for a unified theory of quantum gravity. At its most basic, this interplay between time as a construct and time as an intrinsic feature of reality speaks to our very approach toward the nature of spacetime, causality, and the structure of the universe itself. Thus, the philosophical questions regarding time become essential as we continue on into the mysteries of quantum mechanics and general relativity in a way that may revolutionize the very conception of reality for all of us.

8: WHO OBSERVES THE OBSERVER?

Our discussion has delved deep into both the philosophical and scientific dimensions of the quantum world. As I mentioned in the context of the double-slit experiment, particles exhibit different behaviours when they are observed. This raises a rather interesting question: does this mean that particles somehow "know" when they are being observed? The answer is more complex than just attributing awareness to them. In reality, it's not that they consciously respond to our observation, but rather, the act of observation itself plays a crucial role in determining their behaviour. This brings us to an even more profound question: who is observing us, and what determines our existence in this vast universe?

One theory is that we might be living in a simulation of the world, created by superadvanced beings. But that theory doesn't provide a final explanation. Let's just assume another point of view on our existence. Now consider the question of how something can be made. The most obvious answer is that something must make something. Nothing can come from nothing. This logical principle points to the fact that everything created must have a cause, a reason behind its formation. So, when we turn our attention to the Big Bang—

BEYOND THE ATOMIC LEVEL

the event that initiated the expansion of our universe—we must ask: what caused it? And if we ask what caused that cause, the chain of questions seems endless. Each cause requires a prior cause, suggesting a deeper, recursive pattern to existence.

However, if we take such inquiry further, we get an impasse: since it appears that everything must needs something to bring it into being then seems to be no really true beginning. Maybe that chain of causality doesn't have a headstart, and so is effectively infinite. This is just one way to realize why the notion of such beginning is an illusion. Instead, the universe could belong to an eternal cycle, creation and destruction only phases in an ongoing process. The end result is the notion of thermal death, wherein the universe has reached a state of maximum entropy; all types of matter and energy would lose their different characteristics, becoming uniform heat. All life and activity stop, and the universe enters a state where everything becomes fixed in a motionless, lifeless balance. So, in this greater scope, perhaps there never was a true "beginning" or "end" but a continuous cycle of time, with the final fate of the universe to be determined by the laws of thermodynamics.

You are standing at the edge of a great forest; each path you take takes you to a different place. You, the traveler, can walk in any direction. However, whichever direction you pick creates a reality. On one reality, you're walking through the meadow. You climb a mountain on another reality. In this game of possibilities, every action spawns its corresponding reality so that every outcome has somewhere in the infinity to become a reality. This analogy captures the essence of the Many-Worlds Interpretation (MWI) of quantum mechanics, a theory that suggests every possible outcome of a quantum event occurs in a separate, parallel universe.

The core of quantum mechanics is the phenomenon of

superposition, where a quantum system, such as an electron, exists in multiple states simultaneously. Consider the famous double-slit experiment, where a particle behaves like a wave and passes through two slits at once, creating an interference pattern on the screen. However, when an observer measures the particle's position, it "collapses" into a definite state, behaving as a particle. This collapse, a central feature of the traditional Copenhagen interpretation, suggests that reality does not exist in a definite state until it is observed.

However, this introduces profound questions: What then happens to the other possibilities, which don't collapse? Do they disappear? Or exist in some form? That is precisely what the many-worlds interpretation elegantly, albeit controversially solves.

Proposed by physicist Hugh Everett in 1957, the Many-Worlds Interpretation posits that instead of a single reality collapsing into one definite outcome, every possible outcome of a quantum event occurs—each in its own separate universe. This interpretation removes the need for wave function collapse, suggesting that the universe "splits" into multiple branches, each representing a different outcome. In the case of the double-slit experiment, instead of the wave function of the particle collapsing to one path, it splits into two different realities-one where the particle passed through the first slit and another where it passed through the second.

Every quantum decision, from the position of electrons to the movement of stars, creates a new universe. These universes are not abstract possibilities but real-world outcomes, existing independently in separate non-interacting branches. According to MWI, the universe we experience is just one of many possible realities, each formed by a different outcome of quantum events.

The Many-Worlds Interpretation completely changes our understanding of reality. In a universe governed by MWI, all possible outcomes are realized, so there is no need for the "randomness" associated with quantum mechanics in the traditional sense. Instead, the apparent unpredictability of quantum events is simply a reflection of our limited perspective in one of the many branches of the multiverse. This interpretation further implies that, every time a quantum event occurs with more than one possible outcome, the universe branches into many parallel universes. When you choose between turning left or turning right, one version of you takes one route and the other version takes another route. Both choices are equally real but they exist in two different realities that can never communicate.

One of the striking implications of the Many-Worlds Interpretation is the concept of determinism. In a way, everything that could possibly happen has already happened in some universe. This view challenges the conventional understanding of free will and raises questions about the nature of our decisions and actions. If every possible outcome occurs, then are we truly making choices, or are we merely traveling down one path among many?

Despite its fascinating implications, the Many-Worlds Interpretation has its critics. One of the main objections is that it lacks empirical evidence. Since the parallel universes proposed by MWI do not interact with each other, they remain fundamentally unobservable. This makes it difficult, if not impossible, to test the theory through direct experimentation or observation. The sheer extravagance of the idea—the existence of an infinite number of parallel universes—has led some to question its plausibility and whether it is truly a scientific theory or a metaphysical one.

Third, there seems to be a challenge regarding the concept of a reality - only one, coherent universe. The Many-Worlds Interpretation leads people to ask whether these created universes are distinct entities from each other or just split fragments of a larger unified system. While the theory indeed sidestepped the problem of wave function collapse, it brought new conundrums in respect of the nature of being, identity, and of a concrete notion of reality.

However, de Broglie's interpretation was different from the conventional Copenhagen view, which stresses the wavefunction and probabilities.

In his original proposal, de Broglie suggested that particles do not merely behave probabilistically, but are guided by a "pilot wave" that directs their motion in a deterministic manner. The particle's trajectory is determined by this guiding wave, which evolves according to the Schrödinger equation. Despite being groundbreaking, de Broglie's idea was largely overlooked for many years.

The theory was actually revived by physicist David Bohm in the 1950s and expanded on the work de Broglie had undertaken, which was developed into what is commonly known as the De Broglie-Bohm theory or pilot-wave theory. In this theory, Bohm made available to particles a kind of "quantum potential" that drives their motion in a way that doesn't have the restrictions inherent in the Copenhagen interpretation-that is, its probabilistic outcomes. According to Bohm's variant, particles at any instance possess definite positions and momenta; their dynamics are a sort of stream function following the wavefunction.

In the De Broglie-Bohm interpretation, the quantum systems are described by a wavefunction and a trajectory of the particle. Like in the conventional quantum mechanics, the wavefunction gives the probability distribution of the particle as a whole. However, instead of having just probabilities, it is a guiding field for the particle's motion. The trajectory of the particle then determines the particle's motion due to the influence of a quantum potential derived from the wavefunction.

This theory contrasts with the Copenhagen interpretation, which emphasizes indeterminacy and the probabilistic nature of quantum phenomena. In the pilot-wave theory, there is no built-in randomness; every particle's trajectory is determined even if we can't sometimes measure it exactly. Probabilistic results are due to our lack of knowledge regarding the initial conditions of the system rather than to actual indeterminism.

The central concept of the De Broglie-Bohm theory is the quantum potential, which is a function of the wavefunction. This means that the motion of the particle is influenced by the quantum potential in a non-local manner. In other words, particles can be instantaneously affected by the global properties of the wavefunction, even if they are spatially separated. This nonlocality, which was later associated with quantum entanglement, suggests that the theory provides a more intuitive framework for understanding quantum phenomena.

The quantum potential allows the pilot-wave theory to explain phenomena such as interference and superposition without invoking wavefunction collapse, which is a hallmark of the Copenhagen interpretation. For example, in a double-slit experiment, the particle follows a well-defined path, but its motion is influenced by the interference pattern created by the wavefunction. This gives rise to the observed interference pattern, even though the particle itself is always following a definite trajectory.

The De Broglie-Bohm theory has often been scrutinized in light of Bell's theorem, which deals with the nature of quantum entanglement and local hidden variable theories. Bell's theorem showed that no local hidden variable theory can reproduce all the predictions of quantum mechanics. This was interpreted as a strong argument against deterministic interpretations like Bohm's theory.

However, the De Broglie-Bohm theory does not adhere to the "locality" requirement because it is intrinsically nonlocal. The particles can instantaneously be affected by changes in the wavefunction, independent of distance. This is a nonlocal property of the theory, and as such, though the theory maintains determinism, it cannot be said to abide by Bell's local realism. Nevertheless, the theory has been appreciated for providing an easier, more intuitive visualization of quantum mechanics, without paradoxes of wavefunction collapse.

One of the most attractive features of the De Broglie-Bohm theory is that it allows for a deterministic and causal picture of the quantum world, as opposed to the intrinsic randomness of the Copenhagen interpretation. This is very appealing to those who wish for a more classical view of the universe, in which cause and effect are preserved at all scales.

Yet the pilot-wave theory has not escaped the criticisms. The most striking feature of nonlocality has been an intense point of controversy; in that it suggests the effect from particles on one another in instantaneous ways, challenging the common understanding of space and time. Moreover, a necessary requirement for us always to have access to wavefunctions, which is abstract, mathematical objects that most feel diminishes their physical content.

Furthermore, although the theory is deterministic, it makes no predictions that are different from standard quantum mechanics in scenarios where the wavefunction is completely known. In essence, the De Broglie-Bohm theory is mathematically identical to traditional quantum mechanics for most practical purposes and it does not give a set of experimental results that differ from the standard quantum mechanics.

Some have even theorized that perhaps the universe does not merely consist of all these particles and forces; it is, in a certain sense, co-creative with consciousness. Thereby, it could imply that our consciousness doesn't just perceive reality; instead, it shapes reality actively. If consciousness were never to be, then presumably the universe would be just another state of uncertainty that lies in wait for something capable of focusing it to actuality.

BEYOND THE ATOMIC LEVEL

On the other hand, the idea of an ultimate observer raises some very tough questions about reality. If consciousness causes the wave function to collapse, then perhaps reality is subjective and formed by our minds, or perhaps there is an objective reality independent of observation, independent of whether we are conscious of it or not? If the universe depends on an observer, then what happens when there is no observer at all? Does the universe "wait" for consciousness to perceive it?

Though this question remains unresolved, the one thing that is known is that consciousness plays a vital role in how we experience the universe. Whether it is the quantum collapse of possibilities or the simple act of looking at a distant star, our perception shapes the world around us. Perhaps the most profound answer lies not in whether the universe needs an ultimate observer but in the fact that consciousness and reality are intertwined in a dance of creation, each influencing the other in ways we are only beginning to understand.

These concepts are tightly linked to the mathematical structure of Hilbert spaces, that must be used to describe a state of a system and define the physically measured quantities.

A Hilbert space is a complete vector space equipped with an inner product. Essentially, it's a space for vectors (representing quantum states) and scalars by which such vectors can be multiplied along with the ability to measure, through the inner product, the "angle" between vectors. Vectors of a Hilbert space thus represent the possible states of a quantum system in quantum mechanics.

The key feature of a Hilbert space is its completeness, meaning that any Cauchy sequence (a sequence of vectors that should converge) within the space has a limit that is also in the space. This property ensures that Hilbert spaces provide a solid mathematical foundation for describing quantum states and operations on those states.

In quantum mechanics, a vector in a Hilbert space describes the state of the physical system, typically denoted as $|\psi\rangle$. The vector encompasses all the information regarding properties of the system, like its position, momentum, and energy. Often, $|\psi\rangle$ is referred to as a state vector or wavefunction when it's represented in a particular basis, like the position or momentum space.

An observable, such as position, momentum, or energy, is represented by a linear operator acting on the Hilbert space. These operators correspond to physical quantities that can be measured in an experiment. For instance, the position operator acts on the wavefunction to extract information about the particle's position, while the momentum operator does the same for momentum.

Hilbert spaces are crucially related to the concept of observables in the definition of measurable quantities. The observable in quantum mechanics corresponds to a self-adjoint operator in the Hilbert space; that is, the operator equals its Hermitian conjugate. These operators possess special properties that allow them to be associated with certain values, known as eigenvalues, which are the possible outcomes of measurement of the observable.

When an observable is measured, the system collapses into an eigenstate of the corresponding operator, and the result of the measurement is the associated eigenvalue. Mathematically, this is expressed by the equation:

$A|\psi\rangle = a|\psi\rangle$

Where A is the observable operator, $|\psi\rangle$ is the quantum state, and a is the eigenvalue (the measurement result). The set of all possible eigenvalues of an operator corresponds to the possible outcomes of measuring the observable.

The wavefunction (or state vector) of a quantum system encodes all the possible outcomes of measurements of observables. For example, in position space, the wavefunction gives the probability amplitude of finding a particle at a specific position. When we measure an observable, the wavefunction collapses into one of the eigenstates of the operator, and the measurement result corresponds to the associated eigenvalue.

In this context, Hilbert spaces provide the necessary structure for defining and working with these wavefunctions. Using the inner product of Hilbert space, we are able to calculate probabilities of measurement outcomes since the probability of measuring a particular eigenvalue is given by the square of the inner product between the state vector and the eigenstate associated with that eigenvalue.

The measurement problem in quantum mechanics refers to the puzzle of how and why a quantum system's state collapses upon observation. In Hilbert space, this is represented by the process of projecting the state vector onto an eigenstate of the observable operator. However, the physical interpretation of this process remains a subject of debate, especially considering that the state vector itself evolves deterministically according to the Schrödinger equation, except during measurements.

In this regard, Hilbert spaces offer the mathematical formalism for understanding how quantum states evolve and how observables can be defined but also raise fundamental questions on what constitutes a "measurement" in quantum mechanics.

The fact that we can perceive and measure only some things does not prevent the possibility that there are also fundamental truths or dimensions beyond the reach of our cognition. This notion has it that the universe may be so vast and complex that it can harbor phenomena that are beyond the scope of our senses or even the most advanced scientific tools. This notion goes to challenge the idea that we may ever fully understand reality. It suggests a boundary, where knowledge meets its limits. It questions whether some aspects of existence are intrinsically unknowable, remaining behind the veil of our finite perception.

Now imagine the mind as a huge library. Each book represents a piece of information, and the connections between the books—how one idea links to another—determine the overall structure of the library. But what if, hidden in the depths of this library, there were secret, invisible threads connecting books in ways we couldn't perceive? These threads could represent the mysterious quantum processes that might underlie consciousness.

Imagine the mind as a tremendous library, where each book symbolizes a piece of information and its relationships with other books define it as a library. Some ties are visible, representing links of logic we understand or believe in, while others cannot be seen, like magic threads that bind seemingly disjoint ideas. These hidden threads might be considered the mystery of the human consciousness, a fascinating mystery that has been tackled for centuries by scientists and philosophers. Among the many theories in search of understanding, two have gained considerable importance: Integrated Information Theory (IIT) and the Orchestrated Objective Reduction (Orch-OR) hypothesis.

IIT suggests that consciousness is the result of integration in a system. It underlines that it is the richness of interlinked information that determines the grade of consciousness. For instance, though data is processed by a smartphone, due to lack of integration, it does not attain consciousness. The human brain, however, integrates information in such a complex manner that it produces subjective experience. Although IIT primarily works within classical physics, some interesting analogies to quantum mechanics exist, such as interconnectedness in quantum entanglement. Orch-OR, however, binds consciousness to quantum mechanics head-on. The theory hypothesizes that quantum computations take place within microtubules-the tiny structures within neurons. Under certain conditions, according to Orch-OR, the quantum states collapse and produce moments of conscious awareness. While IIT believes that consciousness is just integrated information, Orch-OR holds that consciousness springs from the very properties of the quantum world, such as coherence and entanglement.

Both theories have their own insights but are problematic. IIT fails to explain the subjective nature of consciousness, often called the "hard problem," while Orch-OR lacks experimental evidence for quantum coherence in biological systems at room temperature. Despite these difficulties, the theories complement each other in some ways. IIT offers a way for understanding consciousness through classical process, but Orch-OR goes a step further down into the quantum realm and predicts that consciousness could be a fundamental property in the universe. Are we only classical information processors or does our consciousness emerge from the quantum weave of reality? IIT and Orch-OR drive us to look beyond the edge of what we think we know. Analogously, in the case of the library, in the search for consciousness, perhaps we must understand not only the apparent order of the books on the shelves but also the quantum connections between them all. This speaks to one of the most fundamental questions in existence: do we, by observing, create the reality we find ourselves in? In quantum mechanics, observation collapses a wavefunction into a definite state, suggesting the reality may depend on whether an observer is present. Are we not participants in this universe but its architects as well?

Consider this: without observation, quantum particles exist as probabilities, mere possibilities floating in an undefined state. Only when measured or observed do they take on a concrete form. If such a principle extended beyond the quantum scale, then it could be said that every tree, star, or thought might be contingent upon being perceived.

BEYOND THE ATOMIC LEVEL

But who or what is the ultimate observer? If our perceptions shape the world around us, do we co-create existence? Or do we just reveal what exists? The idea throws the responsibility of the observer into the hands of consciousness itself, where reality is being actively sculpted. Maybe, existence is not a prewritten story but a canvas constantly painted by our perceptions, choices, and awareness.

But this thought also invites humility. If we are architects, we are limited architects, bounded by the resources of our senses and the frontiers of our minds. Reality may extend far beyond what we see, shaped not just by individual observers but by a joint play of consciousnesses of all kinds—human and non-human, cosmic, and so on. Within this play lies the mystery of existence: a universe seen and seeing, creating and being created.

The idea of being as an endless possibility of existence is so profound and fascinating.

Think of a situation for one second: I give a white sock to a friend.

It is simply an ordinary object, it is only white, non-pretentious, and one. Nevertheless, when my friend puts on the sock, perhaps they ask me, "What colour is this sock? " This simple question throws open the gate to investigate the very concept of reality further. It doesn't have to be. Colour of the sock, as with our reality, isn't absolute truth but just one of the possibilities. And if there was a universe where the sock was red, another universe where it was blue, and another universe where it was green? All these possibilities exist together, waiting to be fulfilled, just like quantum particles that exist in superposition, embodying many states until observed. This analogy serves to illustrate a broader idea that our reality is not limited to being a single, fixed reality but rather unfolds as a dynamic series of potential outcomes each branching out and coexisting in a vast, interconnected web of possibilities. Just as my friend's question about the sock's colour reveals that the answer depends on the universe in which it is observed, so too are our lives shaped by the observer. Each moment is not a predetermined point along a singular path, but a decision made among countless possibilities. The moment we "observe" or make a choice, a specific possibility is collapsed into reality, and the others—though still existing in some potential form—become inaccessible in that moment.

In this way, we are not static beings but projections of possibilities. Each choice we make, each thought we have, and each action we take is a manifestation of a potential outcome. Our consciousness acts as the force that collapses the many possibilities into the singular, observable reality we experience. This concept closely parallels the quantum mechanical principle of the existence of particles in superposition, that they exist in all possible states at once and collapse into a specific state only when measured or observed. Just like how the colour of the sock is uncertain until observed, so too is our reality a vast collection of probabilities all existing together until we observe them, shaping them with our consciousness.

If the analogy is to be carried further, imagine that each observer, or consciousness, is like an individual in his own universe, each observing the sock and attributing a colour to it. One friend might see it as red, another as green, and yet another as blue. These interpretations are not wrong; they simply reflect the different ways reality can manifest depending on the observer. In a multiverse, each observer may experience a different version of the same reality, yet each version is equally valid and real within its own universe. This idea aligns with the concept of the Many-Worlds Interpretation of quantum mechanics, which suggests that all possible outcomes of a quantum event occur, each in a separate universe.

BEYOND THE ATOMIC LEVEL

Thus, in a universe where the colour of the sock is determined only when observed, we are reminded that reality itself is not a singular, static experience but rather a dynamic and ever-changing collection of possibilities brought into existence by our perception and choice. Just as the colour of the sock depends on the observer, so does our experience of reality. Every decision, every action, every thought is not only a passive reaction to the world but an active force that shapes the world itself.

In this framework, reality is not something we discover but something we create. We are not mere bystanders in the universe; we are its architects. Our consciousness is the lens through which infinite possibilities are brought into being. The act of observation—whether that's the colour of a sock, the way of our lives, or the nature of the universe—gives shape to what is now a field of possibilities. In this sense, we are not merely dwelling in a world; rather, we are dwelling in a world we help shape with every observation, every decision, and every choice.

But who or what is the final observer of this vast web of possibility? Are we ourselves co-creators, helping in bringing this world into existence, or simply an unwrapping of something already there? Maybe the answer is a mix of the two. We are both observers and participants in the creation of reality, and though our consciousness shapes the possibilities we experience, there may be a greater force or structure that governs the underlying rules of the multiverse. Perhaps our consciousness is part of a larger cosmic web of awareness, where all observers contribute to the unfolding of existence, each bringing forth their own projections of reality.

If our consciousness shapes reality through observation, do we, as observers, create the very world we inhabit?

This is further similar to the idealism of George Berkeley who asserts that existence is fundamentally based on perception and says, "to be is to be perceived." According to this idea, the physical world does not comprise matter; instead, it is an extension of our perceptions. Quantum mechanics may be considered as supporting idealism because it focuses more on the role of an observer in making a reality.

Yet, even if our perception is the key to shaping reality, what does this mean for our identity? If reality is an ever-shifting projection, then how do we define ourselves in this fluid, uncertain existence? The notion of selfhood becomes equally elusive. Quantum mechanics goes against this idea of a fixed, unified self. Perhaps our identity is not one but many, or even all, possibilities shifting and entangled with the world. Just as a quantum particle does not have a definite position until measured, perhaps our sense of self does not exist in a fixed form but is instead a fluid spectrum of potential identities, collapsing into a singular experience only when observed.

This brings us to the nature of existence itself. If we are indeed projections of possibilities, as quantum theory might suggest, what is the source of our existence? In other words, existence was often understood in traditional philosophy as some kind of given: simply "is." But in the world of quantum phenomena, everything is fluid; even the most concrete concepts such as time and space and matter are themselves not fixed. The world we experience is not a solid, stable structure but an ever-changing set of possibilities, each dependent on observation, interaction, and perception.

As we continue to push the boundaries of our perception, we may realize that the universe is not a static, deterministic system but rather a dynamic and uncertain flow of possibilities, each shaped by the act of observation. The question then becomes not whether the universe is real but how we define "real" in the first place. Perhaps reality is not something that exists independently of us, but something that is co-created in every moment of perception, observation, and interaction.

BEYOND THE ATOMIC LEVEL

In this view, we are not just passive observers of a preexisting reality; we are active participants in the creation of the world around us. Our consciousness and our ability to observe shape the very fabric of existence, turning possibilities into tangible experiences. The world is not just a place we inhabit; it is a place we create through the act of perceiving.

The more we reflect on our influence on the universe through observation, the more we realize that our actions, thoughts, and perceptions are not isolated. They are interwoven with the fabric of existence.

In quantum theory, the observer plays a crucial role in collapsing the wavefunction, determining the state of a system. But this phenomenon doesn't exist in isolation; it mirrors how we interact with our reality. When we view the world through a certain lens-be it with fear, hope, joy, or indifference-we are, in a way, collapsing the potential outcomes of our lives, shaping the trajectory of our personal experience. Our consciousness, in this sense, is not an inert entity; it is an active participant in the unfolding of our existence.

This perspective blurs the boundary between the external and the internal. The act of observation is not merely a perception of the outside world; it is equally about how we engage with our inner world. Just as the quantum observer influences the system they observe, our perceptions and inner states influence the world we engage with. What we focus on, how we interpret our experiences, and the meaning we derive from them all serve to shape our personal reality.

It is a reflection on the philosophy that brings us back to the core question: Who are we, the observers? If we are not just passive witnesses to the unfolding universe, then what does this say about the nature of consciousness itself? Are we simply products of the universe, or are we, in some way, co-creators? This question reverberates down the corridors of science and philosophy, suggesting that perhaps we might be over-imaginative and consciousness can be much more complicated than what we have thought.

Thus, before we conclude, we must remember that observation is not just a scientific fact-it's also a philosophical and spiritual one. Our position as observers is much deeper than what we might have ever understood. It connects us not only to the world we see but to the very fabric of reality itself, making us both witnesses and participants in the creation of the universe. The act of observation is, in itself, a reflection of our consciousness, and in this reflection, we find the key to understanding our place in the cosmos.

One more philosophical concept that gets enmeshed with observation and consciousness is the idea of ontological idealism. Ontological idealism speaks to the nature of reality: specifically, that the nature that we experience in this existence is fundamentally mental or spiritual by its very nature. So then, whereas materialism finds the physical world as independently created outside of perception, this latter concept suggests that the nature itself-ontology-is mind-generated. This means that the universe doesn't just exist "out there," waiting to be perceived; it is perpetually constructed by our consciousness.

If we apply this perspective to quantum mechanics, then we could argue that it is not merely that some physical system is changed due to an external observer's influence. Instead, the observer's consciousness may well be considered a player that actively participates in the construction of that reality, much like the idealists suggest. The universe is perhaps not "real" as one may consider it but rather an image of the mind, and constantly in flux in response to our thoughts and perceptions. This falls into the school of thought of quantum mechanics, in which an observer is considered to determine what the state of a given system is.

Another, more profound challenge, made by this

philosophy is how it deems everything; atoms, particles, or even the cosmos may somehow be founded on consciousness. This sounds similar to ideal observer theories in quantum mechanics, where the reality being created depends on an observer. The concept "consciousness creates reality" is supported when one considers that our perception of the world is not a passive reflection, but an active engagement that shapes the world around us.

Yet, another philosophical approach relates the anthropic principle, wherein, it is believed that the universe appears to be fine-tuned for the existence of life because the conditions for life necessarily have to be compatible with human observation. This principle suggests that the very fact that we are here, capable of observing the universe, may be an essential aspect of how the universe exists. It asks whether the universe is in some way shaped by the fact that we are conscious beings observing it or whether it would exist in the same way regardless of our presence.

These philosophical concepts deepen our understanding of quantum mechanics and reality because observation is not merely an external, passive act but a dynamic and integral part of the unfolding of existence itself. The questions of whether the universe is real, whether it exists independently of consciousness, or whether consciousness itself is part of the fabric of reality challenge us to rethink not just the mechanics of the universe, but the very essence of what it means to exist, to observe, and to understand.

The Anthropic Principle is a cosmological concept that states the universe's physical constants and laws seem to be finely tuned to support the existence of life, especially human life. At its core, it posits that the universe appears to be structured in such a way that it allows for conscious beings to observe and reflect upon it. This idea raises intriguing questions about the very nature of existence. The principle is often invoked to explain why certain aspects of the universe—such as the strengths of gravitational and electromagnetic forces—are precisely calibrated to allow for life to emerge. There are two main interpretations of the anthropic principle: the Weak Anthropic Principle (WAP) and the Strong Anthropic Principle (SAP). The weak version suggests that the universe must have the properties it does simply because, if it didn't, we wouldn't be here to observe it. It's a straightforward explanation that our existence and the universe's properties are intrinsically linked—without the right conditions, life wouldn't be possible, and thus we wouldn't exist to question it. The strong version, however, takes a much more profound stance, implying that the universe's properties are fine-tuned with purpose, or in such a way that the emergence of intelligent life is not just possible but inevitable. The strong anthropic principle suggests a sort of cosmic design where the universe's laws seem to be specifically set up for life to arise. This is an interesting possibility, but it has also been criticized for potentially becoming a tautology: a circular argument where the universe is said to be the way it is simply because it allows for life, with no real explanation for why the universe should be this way in the first place. Critics argue that it doesn't offer new insight but simply restates the obvious. However, for supporters of the anthropic principle, this is a lens by which we must view the universe as it exists uniquely. Proponents feel that life and consciousness cannot be considered accidental but, instead, might reflect deeper connections with the very structure of the universe. In the contexts of observation and consciousness, the anthropic principle demands further philosophical thought. Does the universe, in some sense, have to be the way it is because conscious beings, like us, can perceive and understand it, or is life simply a byproduct of particular physical laws and conditions, without inherent purpose or design? These questions invite further exploration of the fundamental nature of reality, challenging our understanding of existence, the role of the observer, and the potential reasons behind the universe's remarkable capacity to support life.

Who is observing us?

I believe that something in the universe is observing us, not in the traditional sense of a god, but in a way that transcends our understanding. At the quantum level, it seems like there is an underlying presence or force that is aware of every possibility, shaping reality based on the endless potential outcomes. This observer doesn't necessarily need to be conscious in the way we are, but it holds the knowledge of all possible states, influencing our existence in ways we might not fully comprehend. It is as though our reality is constantly being molded by an invisible force that knows and guides each possibility.

As we bring this sojourn into the nature of reality, quantum mechanics, and consciousness to its end, it is fitting that we reflect on our role as observers within the vast tapestry of existence. For most of our lives, we have been trained to regard ourselves as separate from the world surrounding us, passive observers of what transpires. Quantum mechanics challenges this very assumption, for it teaches that observation is not just a neutral, detached occurrence, but an active force shaping the world.

In quantum mechanics, observation is not merely the act of perceiving something; it is the act that brings potentialities into being. The famous double-slit experiment demonstrates that particles can behave as both waves and particles depending on whether or not they are observed. This means that, at a fundamental level, the world exists in a state of uncertainty until we, as observers, intervene. We cannot be mere spectators of the universe; we are, in a sense, co-creators of reality itself.

This realization leads us to a very deep philosophical question: What does it mean to be an observer? In our everyday life, we may consider ourselves as independent entities, existing outside of the universe we view. Yet, as we have seen in the quantum mechanical perspective, lines of distinction between observer and observed become very fuzzy. In observing a quantum system, we do not just discover its state, but in fact define it as we observe it. Our process of observation is what brings change to the system - its wavefunction collapses, with potentiality now being actuality. It's like the universe is waiting on some level for us to relate to it in order to actualize itself.

And reflecting on that gives one pause: Is there really a distinction between us and the world? Or do we, being so thoroughly participating in the reality that appears before us, have some illusions about how separate we really are from this universe. We might be more actively involved in a vast, all-inclusive web of existence, and not isolated beings. We are part of the world and also, through our perception and observation, contributors to its unfolding.

The idea that our consciousness and our acts of observation influence the world has fundamental implications. It suggests that reality is not just an objective set of facts, independent of us. Instead, it is highly intertwined with our consciousness, even with our existence. We are part of the mystery itself, in a very real sense, in that we are not only passive recipients of the mysteries of the universe but also active creators of them. Every decision we make, every measurement we perform, ripples across the fabric of existence, creating and modifying the universe in ways both subtle and profound, potentially beyond our full understanding.

This also makes us ponder over the ethical dimension of our observation. If our observations can influence the reality around us, then how we observe becomes important in every possible way. The act of observation is not a scientific concept in the quantum world; it is philosophical. It is how

BEYOND THE ATOMIC LEVEL

we perceive the world, interact with others, and even ourselves as observers that in their own ways become acts of observation. All interactions, big or small, add up to this enormous web of existence, and all observers are a part of that bigger web.

At a personal level, it makes us reflect on our role as observers and take responsibility for the way we shape our reality. If we are conscious beings who can change the world around us through our perceptions and actions, then what kind of observers do we want to be? Do we want to stand there and watch the world just evolve, or do we want to take an active hand in shaping a reality that shows our highest values and highest aspirations?

We could reflect on how our consciousness, imperfect as it is, becomes the prism through which the universe is interpreted. Our decisions, the way we interpret our experiences, and how we live with the world all leave imprints. The question is not just what we observe but how we observe and what that says about us. Each observation, each thought, each interaction, is a chance to reshape the reality we experience.

Finally, this self-reflection takes us back to the understanding that one is not only a passive onlooker but rather a dynamic and interactive engagement with the cosmos. In this respect, we are at the same time creating and being changed by the world that we are surveying. Our consciousness and our perceptions as well influence the world in ways about which we have just barely begun to grasp. In this respect, we are not separated from the universe but, rather, integral parts of its continuous unfolding, in turn, shaping and being shaped by the reality we engage with.

So, as you look back on the mysteries of the quantum and the nature of observation, remember that you're not just looking in at the universe from outside; you're part of it, helping it create itself. And that's the point-the very act of

BEYOND THE ATOMIC LEVEL

observing in the quantum is transformative. So now it becomes not so much a question of what the universe is but how you are helping to shape it, you, the observer.

The more we try to explore the universe and the mysteries of its vast expanse, the more we encounter questions that put into test our reality. We can look and measure particles at an atomic level, but how far can we stretch our eyes beyond the horizon of what we can physically see? Is it the same with the universe, or does it play by rules and patterns different from those we are accustomed to? As we go about defining matter, consciousness, and time, we are bound to ask ourselves whether our senses and tools are sufficient to apprehend the essence of things or if there is something much more complex and hidden, far beyond our observations. Do the answers lie in what we can see, or do they lie somewhere beyond the atomic level?.

BEYOND THE ATOMIC LEVEL

ABOUT THE AUTHOR

Aryan Vinod is an Indian author and aspiring short filmmaker, born on February 12, 2007, in Idukki, Kerala, India. He specializes in writing detective and mystery stories, known for their intricate narrative structures and compelling suspense. In addition to his fiction work, Aryan has a keen interest in science and has authored books on the subject. His passion for storytelling extends to filmmaking, where he is working towards bringing his creative ideas to life through short films